A STORY ABOUT
NOTHING

A Story About Nothing

or How Things Came to Be

Patrick Hentsch

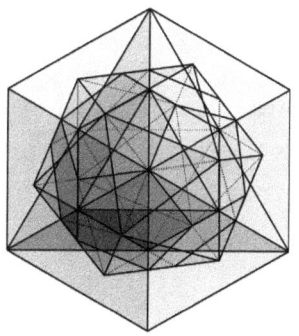

ISBN: 9798379121518

To Felix, Elektra, Daphne, and Penelope,
for being everything to me.

TABLE OF CONTENTS

A Story About Nothing

Chapter 1: Once Upon a Time

I would start my tale with the customary "*Once upon a time...*" if it weren't for the fact that this particular story begins before any other beginning had begun; before time had spent enough of itself to know of its own existence; indeed, before there was any kind of time once upon which this story could have announced its opening.

In the Beginning There Was Nothing

Strangely—you see—the beginning is a difficult time or place to get anything started. One might reasonably have thought the beginning to be a most natural place to begin. But all those beginnings, to which we are so accustomed, rely upon countless other beginnings that all came one before the other. But an absolute beginning has nothing that precedes it, and so nothing is in place. There are not even places in which for anything to be, let alone things that need any place in which to be. But I did say that "nothing was in place" and I meant it *almost* literally: *almost* instead of completely literally, because I also just told you there were no places yet; but that didn't prevent nothing from already being *in place*, which after all cannot mean quite the same thing as being *in a place*. That's the cleverest thing about nothing: Nothing—not being anything—needs neither moment nor place in which to be accommodated. Nothing is the only thing that cannot be preceded by anything else, not even by time nor by space, which is the same as saying that *nothing* can precede nothing. If you've paid careful attention to the previous sentence, you may notice what it really says, which is that *nothing* is the *only* thing that *can* precede itself, which alone is a fascinating enigma[1]. So when I said that "nothing was in place" what I meant was that nothing is far from no-thing: it is a very

---

1. *Specifically, this is the enigma of self-referential paradox. Many of the most bewildering phenomena in the universe are self-referentially paradoxical. We take external frames of absolute reference for granted, without which we would be thoroughly disoriented. Our planet Earth is such an example. We use immobile physical features like topography and buildings to ascertain where we are located and where we are going, but as far as the solar system is concerned my bedroom is constantly changing position relative to the Sun. As far as any other planet is concerned I am in a completely different location each time I return to the same place in my bedroom.*

The concepts in Einstein's theories of Relativity that are the most perplexing are so because they implicitly acknowledge that space, time, (spacetime) and the speed of light (space ÷ time) have only each other or itself (spacetime) with which to define and measure itself and "each other".

A more familiar and easily comprehensible encounter with self-referential paradox is found in statements such as "everything I state is untrue": if my statement is taken to be true, then its own message declares it to be false; if the statement is accepted as untrue, then it validates that the truth is contrary to its own message, that is, the statement is true. However we cycle through this endless loop of simultaneously valid opposite assertions (paradox), the statement makes pronouncements about itself (self-referential), and the result is that there is no static truth or point of reference—unlike what we are used to having in ordinary daily life.

special thing indeed. It is the only thing that can exist without being held by time nor held by space, which means that it must be eternal and infinite. And so the original beginning began with nothing, around nothing, inside of nothing... However we put it, nothing preceded the beginning, and so the beginning had only nothing with which to start, because there was nothing else[2]. As I warned, nothing is deceptively clever; it finds ways to be everywhere all the time without being anywhere ever, and this includes being there before the beginning, and therefore at the beginning, where only nothing—and nothing else—could have been.

A Difficult Story to Tell

Of course my story is very difficult to tell, because I'm using so many things—which include time, space, and words—that weren't themselves there at the beginning. It's rather marvelous, if not ironic, how much we take beginnings for granted. We're so certain that beginnings start at the beginning. We count on it. And yet there are no times nor places so murky, so inarticulate as

---

2. Incidentally, *Nothing else* is just another name for nothing. It's for when nothing wants to be two or more nothings instead of just one nothing. *Nothing else* is all the other nothings, which of course amount to nothing at all...

beginnings. Think of the beginning of your own life. So murky, so inarticulate, that whatever things had to be in place for memories to form were all missing—because it was the beginning and there was a lot of *nothing else* around, and not as many things[3] as you are able to distinguish today. And so when—as an adult—I try to tell the story of the beginning of my own existence, I find myself cheating rather a lot. That is to say, I use things that weren't part of the story to tell the story. If I'm not careful, I may borrow other people's stories about the beginning of my own life and mistake it as my own—which seems a rather frightening idea. One wonders how true a story can be, which is being told with things that were not in existence at the time of the story itself. Surely such a story reveals far more about the present than about the mysteriously simple past that it is trying to describe? And so I've hardly begun this story, and I find that I've already cheated a lot. Even telling the story of nothing preceding the beginning is to have cheated. But I also know that I can't try to tell this story at all unless I cheat, and so I've decided to cheat as

---

3. *As we develop from infancy into maturity, the distinctions we make in our perceptions and observations grow in number. This is reflected by the growth of our vocabulary, and as we assign names (symbolic representations) to our perceptions and observations, they become things (objects) that emerge into existence in the field of our consciousness (subject).*

little as possible, and at the same time to cheat as much as necessary. From those two facts alone, I am certain that some of my readers will strongly disagree with my telling of this story, and others will agree (even though it's nothing but a story, or a story about nothing). But this is not a story told for agreement or for disagreement. It is a tale told for marvel, enchantment, and delight. It wants to be told at night, around a fire, under the stars and moonlight. It wants to imperceptibly bridge the chasm between the reasonable world of daylight and the enigmatic world of our dreams. It wants the listener to remember a way of knowing that later on got swept away by too much information. So let me try again.

NOTHING IS NOT NOTHING

In the beginning, before any beginning had ever begun, there was the minimum necessary for a beginning to be able to occur. And that minimum was nothing. It was so minimum that nothing didn't look like it was any thing at all. But it was something. It was quite something indeed. Perhaps that's why we've since given the most important only original thing—nothing—the most imaginably insignificant name: Zero. Nobody looks at Zero. It's the one thing that can be there without being anywhere. It fills the empty spaces. In fact, it *is* the empty spaces. It's even what holds the empty spaces, and anything that can fill those. Well, Zero had a sense that it was deeply mysterious, and peculiarly powerful. And it also found itself terribly lonely. It had very good reason to doubt its own existence, which may be why it came up with the idea that since there was only nothing, there could be nothing else. Now at least, Zero could start something playful, which is what we like to do when we are lonely and bored. So nothing tried to add nothing else to itself, and discovered that it ended up right where it already was: with nothing ($0 + 0 = 0$). That was disappointing, so nothing tried to multiply itself by nothing else, but it continued to end up with nothing ($0 \times 0 = 0$) and its loneliness continued. When we want *more things*, we add to, or multiply them to end up with more than we started out, but we never wonder where those *more things* came from. But such was not to be the case for Zero, and Zero was very involved in trying to answer the question of where the *more things* were going to come from, when there was only nothing to begin with and nothing else. As Zero struggled to come up with *more* than nothing, it also came to realize that there couldn't be *less* than nothing: there was nothing to be less than, which it already itself was. It had eternally been, so how could it be removed? Where would it be taken away to? It became increasingly curious and, in its loneliness and boredom, Zero got bolder and nothing decided to risk subtracting nothing else from itself. However, it found yet again that the result was no different ($0 - 0 = 0$), but at least now Zero was emboldened by the discovery of its own indestructibility.

DIVISION BY ZERO

We will never know how long this went on for, because without any change happening, time cannot be measured. But it doesn't matter because time and space hadn't even shown up until Zero made its audacious breakthrough. Invigorated by the discovery of its own immortality, nothing now came to the idea of dividing itself by nothing else. Now, unlike addition, subtraction, and multiplication in which Zero had only to attempt certain combinations of nothing with nothing else, division was going to mean splitting and shattering nothing up with nothing else, so this was the most daring venture yet: indeed, Zero had nothing to lose, which in this very special case was everything. And so nothing went ahead and divided itself by nothing else. And what happened next was simply astonishing. Instantly. Explosively. Zero's loneliness and boredom were annihilated and all of a sudden Zero stood face to face with its mysteriously revealed overwhelming twin, Infinity $(0 \div 0 = \infty)$[4]. Instead of losing nothing—which to Zero would have been everything—Zero was now suddenly actually in the company of everything, which meant that for the very first time Zero came into its own company. Since nothing was already everything, everything had always already been there, but now nothing and everything could see each other as two separated things, instead of one and the same. Through division, Zero had transformed itself (0) from

---

4. *Now, I'm aware that it's exactly at moments like this one in my story that some of my readers—maybe in particular those who've received so much academic education that it's become easier to trust rules than it is to sense truth—may disagree. They would, not incorrectly, remind me that we were taught that in ordinary arithmetic, dividing by zero is undefined, and is not therefore infinity. But when nothing was playing with nothing else prior to all beginnings, even ordinariness did not exist. And indeed both Zero and Infinity continue to this day to be anything but ordinary. They are the twinned mystery that revealed everything else, including ordinariness, which came after them. And without them—or the immense space they opened up between each other—all the numbers with which we are willing to perform ordinary arithmetics wouldn't even exist.*

Humans, who sometimes think too much and in wrong ways, correctly argue that zero and infinity are not numbers, and so they become afraid of performing important mathematical operations with these incomprehensible mysteries. Thus they find comfort in declaring that division by zero is undefined, which is ironically and correctly another way of saying that the rule about division by zero is that there cannot be a rule. And this humbly acknowledges the mysterious truth that both Zero and Infinity exist beyond the realm of rules, or numbers, but give rise to all those.

being the one and only (1) nothing to being a twin (2) sibling with Infinity (∞)[5].

---

5. *So let's get over this hump that would prevent us from marveling at how everything began out of nothing:*

We know that multiplication and division are two sides of the same coin. If we start with one thing, but want to turn one thing into four things, we must cut (divide) the thing into four pieces; to increase one thing into four things (multiply one into four) ironically we have to divide by four.

1 thing ÷ 4 = 4 things (that are each ¼ the size of the 1 thing)

Now we can separate (subtract) any of the four pieces to give them away.

And now we can go in the other direction: if we need to turn four things into one thing, we must bring the pieces back (add) and combine (multiply) the four things into one whole thing again. To reduce four things into one thing (divide four into one) ironically we have to multiply by four.

4 things (that are each ¼ the size of the 1 thing) x 4 = 1 thing

So both division and multiplication by the same factor reverse what the other does. The order doesn't matter. If we do one followed by the other, it's like nothing happened. But in terms of original beginnings, division had to happen first, because there was only One nothing, and this is why Zero couldn't get anything out of multiplication, addition, or subtraction before dividing itself into an infinite number of things.

Now, to better understand the mysteriousness of what went on, let's consider the following. Mathematically, what we just looked at can be written as:

1 ÷ 4 = ¼ (also known as 0.25), but what we mustn't overlook is the fact that there are now 4 of these pieces lying around, and size—or proportion—has come into play (0.25 is smaller than 1). And should two of those smaller pieces wish to break away from the group and play games with each other, while ignoring the others—which they now can, having been separated—new mysteries emerge.

If one ¼ tries to divide another ¼, the result is 1 (¼ ÷ ¼ = 1)! And if both pairs of ¼ do that, we end of with 2 ([¼ ÷ ¼] + [¼ ÷ ¼] = 2), which is twice more than we started with! As soon as division occurs, "all hell can break loose."

What we also see is that One, like Zero, is also special and mysterious. Curiously, dividing or multiplying by One itself changes nothing. Dividing by anything smaller than One increases what we started with, which is like multiplying by something correspondingly larger than One (1 ÷ ¼ = 1 x 4).

So if we understand that Zero is the "most possibly smaller" than One, and that Infinity is the "most possibly larger" than One, we would have to understand that division by Zero is the same as

The Subtlety of Nothing Else

The story is subtler in fact, and those of you who have been paying very close attention will already have noted that *nothing* had already conceived its twin the instant it had had the idea of being *nothing else*. Infinity was already hiding right there within the *else*. But it was only through division that Zero's loneliness could be shattered. Zero was looking for a companion, an Other. And although in its longing for company it had unwittingly imagined the Other by having an idea about nothing else, in order to create the Other, nothing had to manifest separation from nothing else through division which meant sharing itself with itself which gave rise to infinite fragments. It was necessary because, until then, no amount of receiving (addition), giving (subtraction), or combining (multiplication) could cure Zero of its loneliness and boredom. It always still ended up with only Zero, or Self: there was no Other. Nothing had to chop itself up (division) with nothing else

---

multiplication by Infinity, and vice versa. And that is what happened when nothing divided itself by nothing else, which is also when it discovered that midway between Zero and Infinity, an equally mysterious and eternal mirror, One, had been waiting to reflect the twins' existences back to each other. Nothing had always been everything, and everything has always been nothing as well. And this could only be done through One. And between each other, Zero, One, and Infinity came to know themselves to be Three, or a Trinity, that would contain and generate the rest of all existence.

in order to bring forth Other. Only now was there an Other to *receive from* (addition), to *give to* (subtraction), and to *combine with* (multiplication), and this brought more meaning and less doubt to the fact of Zero's own existence (consciousness), which was all the benefit. And the price would be the pain of breaking up wholeness into separated fragments, but it was worth it. It's not as if Zero had anything else to experience. It had already been painful to be wholly alone as nothing—as Zero.

THE ARRIVAL OF INFINITY

Infinity's sudden and stunning appearance is where our story really begins, because in the space and time that unfurled between Zero and Infinity, all subsequent stories and their beginnings—both all of those that have already been told, and those that are still being told, as well as the many more that yet remain to be told—became possible.

With this new, expanded view, one of Zero's early realizations about itself was that nothing had always in fact been something, but that until there was something else, something could only know itself as nothing and nothing else. Specifically, that it had been the one and only thing, which awakened it to the revelation that Zero was also One. We humans have subsequently tried to name the

One, variously as *Elohim, Yahweh, Adonai, Jehovah, God, Allah, Brahman, Nirankar, the Big Bang, the Singularity,* amongst countless others... which is ironic because humans have been caught in, and fighting over the tension between One and Infinity, while forgetting about the original, all-too-easily-overlooked, but almighty Zero, as though all three were in fact different and could not be true together. But the Taoists were particularly insightful because instead of trying to name the One, they reluctantly chose to give a name to nothing, to Zero, which preceded all else, and they humbly called it the Tao (道)[6], or "the Great Void".

THE ONE AND ONLY NOTHING

Now we see that Zero being nothing, was the only possible one-and-only thing that existed before any beginning and so, secretly and unknown to itself, Zero was already One. But One was alone ("all-one") and had nowhere to go and nothing to play with. By dividing itself

---

6. *The Taoists also discerned that the One, also known as the Whole (◐), was held in existence by a dynamic equilibrium between the opposite twins Zero and Infinity, empty and full, or Yin (陰) and Yang (陽).*
In their own way, the Christians unwittingly did something similar by giving their God an evil twin, Satan, but in doing so they reassigned God from the One to either Zero or Infinity, depending on how you look at it, leaving the other to Satan, and the ambiguity has continued to confuse us ever since. When everything is as interconnected and as interchangeable as Zero, One and Infinity are with each other, it is easy to lose sight of any comprehensible hierarchy, so it is worth remembering how our story began, which was with nothing.

by Zero[7], it sprang Infinity into being and suddenly Zero and Infinity were thrust to the opposite ends of a creation that opened up between them, expanding away from One, which both remained in the original center, while also expanding with the new, growing perimeter to contain everything that opened in between, within it. But we know that the opposite ends of the Universe, Zero and Infinity, are in fact One and the same, and with some imagination, we might conceive that as they expand away from each other, unfolding a Universe between them, they are also approaching and joining back into One; they are forever moving away and moving towards each other at the same time. They are both always and never touching each other. Where they come together is beyond our known Universe, beyond space and time. A realm in which space and time are both zero and infinite, referred to by some as Eternity, or the Great Void. It is the dimensionless realm that cannot be entered by mass, or matter, but in which all matter is mentally interconnected and from which all matter is spawned. But let us not get ahead of ourselves too quickly, even if the Infinity explosion itself, also

---

7. *We can wonder whether nothing divided Zero by Zero, or One by Zero, but it makes no difference. The result was Infinity, as dividing anything by Zero always produces Infinity.*

described as a Big Bang, appears to have had an instantaneous immediacy[8] to it.

CHAPTER 2: What Immediately Followed

TOO FAST TO TELL

Our story does not become any easier to tell at this point because, while telling about nothing wasn't exactly straightforward, trying to chronicle the multitude of things that flashed into existence upon the Infinity-explosion obscures the fact that these events were synchronous with the instantaneous unfurling of space and time themselves. How are we to tell about anything without the telling—itself—needing to unfold over the passing of time? Language and music, and also stories, are structured against time. And since we humans experience time as a sequential flow of happenings, our stories can only tell themselves as a chain of causes and effects. But what if Nothing, Zero, One, Infinity, Space, Time, and

---

8. *We little, time-and-space-bound humans—as long as we are manifesting as material beings—will always seek to understand things in terms of that which is within the bounds of Zero and Infinity, which is to say of space and time, which includes speed, because speed is simply space divided by time. But we cannot in fact sensibly speculate about the speed at which space and time deployed. Speculations about the speed of speed's own inception is a self-referential measurement, which cannot indicate anything objectively meaningful.*

the Universe all were one another's causes as well as one another's effects—all at once—in a primordial, and also self-referentially paradoxical condition that preceded the possibility of sequence? We, however, are forced to describe happenings in turn, as though they occur sequentially over time.

For example, let's imagine how a tourist's guidebook might describe a castle: Necessarily, the description unfolds as a journey through the fortress and its grounds. The narrator travels from room to room, and from courtyard to garden, and this movement through space extends over time. As words are used to describe what each space holds, the narrator's focus of attention also travels over time through space from one feature to the next, such that while the throne's crimson velvet upholstery is noted, the golden, claw-like feet await their turn to be observed, as do the paintings, the chandeliers, all the furnishings in the adjacent rooms (they will have their turn in the guidebook's pages), the courtyard, and the gardens. For that matter, the narrator may have something to say about the village beyond the moat; And even if the scope of this guidebook ends there, technically the rest of the entire universe is available for attention and description. In that

single moment, everything beside the throne's upholstery
lies outside of the narrator and their reader's beams of
attention. With the assistance of our bodies, it is *attention*
and *description* that must meander through space and across
time to take in and build up a complete awareness of the
entire castle. *Attention* is a traveler that collects things about
which awareness can be had; memory is where the
collection is gathered, sorted and stored. But if that castle
were your home, and you knew it very well because your
awareness of it in its entirety was well-developed, then at
any given instant you could "see" it in its entirety, not with
your eyes, but with the "mind's eye". That ability would be
akin to looking at a doll-house and its contents, all made
of semi-transparent material, such that your perception
and knowledge of the doll-house and its contents could be
complete and immediate. This would be particularly true
had you been the architect and builder of said house.
Unlike the laborious guided tour of the castle that can only
lead a narrow beam of attention from one focus of
awareness to the next, there is a form of awareness that can
be quite expansive, non-linear, and instantaneous.
Architects have it about the buildings they've designed,
and until we understand how we are the architects of our

own lives, we will not have it about our own lives. But it is possible. Through practices of mindfulness we can train the breadth of our beam of attention to grow larger and more inclusive so that we can instantaneously drink-in larger swaths of awareness.

Nothing Began to Play with Infinity

So even if a multitude of things all happened simultaneously at the beginning, because time did not yet exist, we must tell our story as though it unfolded over time, like the guidebook writer. The beginning looked something like this: Zero, after dividing itself by itself, exploded into Infinity which now contained Zero. Between Zero and Infinity, an infinitely large space opened up that would contain absolutely everything. This space was uniform and boring, but pregnant with infinite possibilities. Beside itself, there could be no other space, which makes us wonder what it is that held Infinity itself. But about that all we can do is marvel with awe because there are limits to what mortal human beings can know. To our minds, the space opened up by nothing may have looked like a sphere whose surface is ever-expanding and can never be reached, and whose center was the original Zero. But now suddenly, this space was filled with an infinite number of other zeroes—or points—each of which was the center of another ever-expanding sphere. So although it all looked quite empty at first, the space was in fact quite crowded with infinite overlapping spheres and their centers. It's like having a very large blank sheet of paper. You can place a dot anywhere you like, and then trace a circle of any size you choose. But the sheet is pregnant with all the other circles that you didn't draw all over the page. When nothing decided to focus on one circle—the one around itself—it noticed that a circle was

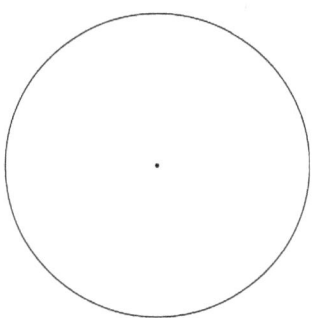

made up of other points just like itself, and that they were all the same distance away from nothing. Nothing was very happy, because it could finally play. It decided to jump over to one of the infinite number of points around itself on the circle; they were all identical and equally distant so it didn't matter which one. Once there, nothing saw the original center, from which it had come, and it decided to expand itself into a second circle whose perimeter would touch the point from which it had come. Nothing saw that

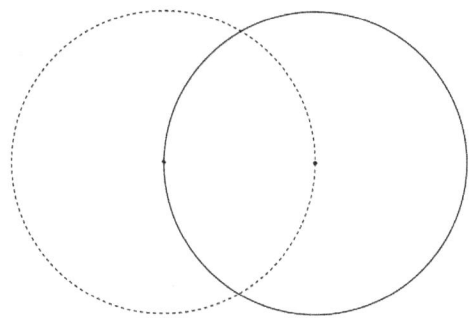

it had created Two out of One. It also noticed that the two circles crossed each other at two points along their perimeters, and given their equivalence, nothing jumped to one of those two points and expanded into a third circle of equal size to touch both previous centers. Out of Two

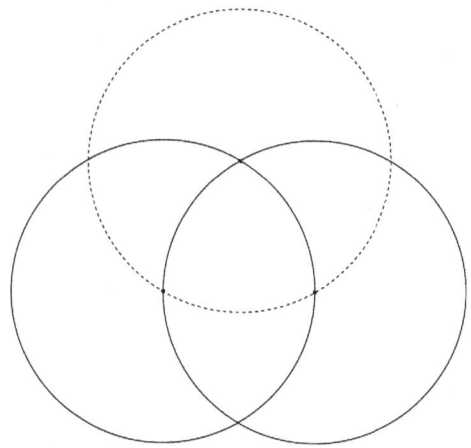

there were Three. Nothing was pleased, and continued to

jump to the next newest crossing along its original circle, forming Four,

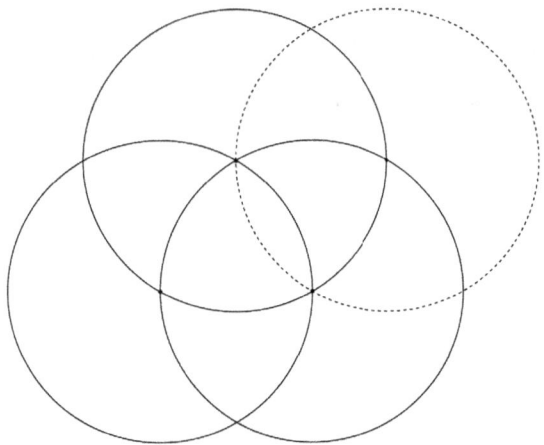

then Five,

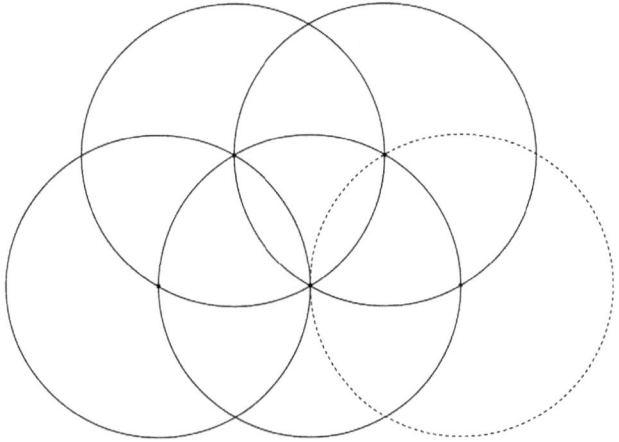

then Six,

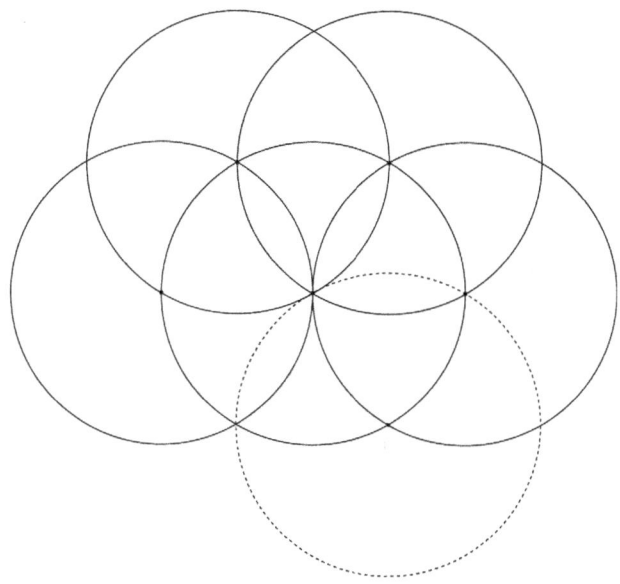

and finally Seven where it realized that it had completed

the circle. Nothing was very pleased with the result and so it paused to enjoy what it had created, which we call *"the Seed of Life"*, but also *"the Seven Days of Creation"*.

THE FLOWER OF LIFE

Nothing saw the beauty and the possibilities in the Seed of Life. Seeds can grow, and nothing saw the six new intersections on the outer circles where six further circles could be grown.

These produced six more new intersections upon which nothing grew another six circles.

This brought pattern and order to the infinite space that had opened up after nothing had divided itself by Zero, and there was no reason to stop, so nothing continued to fill all of space with a continuation of circles at every new intersection.

It noticed at one point of this continuing expansion that the original center circle was perfectly surrounded by six circles, forming once again a beautiful grouping of seven circles. We call this pattern *"the Flower of Life"*. It is *the Seed of Life* expanded.

We often show the *Flower of Life* with an orientation that is rotated by 30 degrees from the previous diagram because it is easier to appreciate the patterns:

If we go further out from the *Flower of Life*, we have what we call "*the Fruit of Life*". There is a logical progression in our names, from seed, to flower, to fruit; each gives rise to the next, each grows out of the previous one.

In *the Fruit of Life*, we find the blueprint for the two-dimensional projections[9] of the five Platonic Solids[10].

---

9. *Two-dimensional projection is also known as a planar projection; it's like taking a three-dimensional object and flattening it into a single sheet (or plane). How the object ends up looking on the sheet is its 2D projection. Not unlike the cartoon characters that were often flattened by steam-rollers into a flat shape.*

10. *The five Platonic solids are the Tetrahedron, the Cube, the Octahedron, the Dodecahedron, and the Icosahedron. They are the only five regular 3D solids, i.e. solids that are enclosed by identical, equal-sided (equilateral) faces.*

The Platonic Solids

But first, let's see how nothing came up with these fundamental solids. By its own expansion, it had already created the perfect sphere whose two-dimensional projection is the circle. From *the Flower of Life* nothing found that after the circle, the most basic flat shape is the equilateral (equal-sided) triangle. The triangle encloses the circle perfectly with three straight lines that join the centers of three of the outer circles. This cannot be done with fewer than three equal-sided lines.

Nothing tried to join two triangles into a solid, but they only folded onto each other into a flat plane. But it went on to find that three, four, and five triangles could be joined around a point to form three different regular solids, and we call those the *tetrahedron*, *octahedron*, and *icosahedron* because they have four, eight, and twenty faces respectively[11]. Joining six triangles around a point simply made a flat six-sided shape that we call the *hexagon*, which could already be found in *the Flower of Life*.

---

11. *In Ancient Greek, tetra- (τετρα-) means four, octa- (ὀκτά-) means eight, and icosa- (εἴκοσι-) means twenty.*

By joining three equilateral triangles around a point, nothing was able to close the solid with a fourth triangle, forming the four-faceted *tetrahedron*.

By joining four triangles around a point, nothing created a four-sided pyramid with an open square at its base; by joining two of these at their square bases, the eight-faceted *octahedron* was formed.

By joining five triangles around a point, nothing created a five-sided pyramid with a pentagonal base; by joining two of those with a row of ten more triangles,

the twenty-faceted *icosahedron* was formed.

Nothing then saw that for each of the three solids, it could join the centers of each face to each other with straight lines.

With the tetrahedron, it found that this created another smaller tetrahedron within it.

Within the octahedron, nothing found a six-faceted solid—technically called the *hexahedron*— that we know as the *cube,* whose faces were squares.

Within the icosahedron, nothing found a twelve-faceted solid called the *dodecahedron* whose faces were pentagons.

By repeating the process of joining the centers of each face to each other with straight lines, nothing found that the tetrahedron kept producing tetrahedra. The cube produced another octahedron within it, and the dodecahedron produced another icosahedron within it.

Thus nothing found that tetrahedra generate tetrahedra, that octahedra and cubes generate each other, and that icosahedra and dodecahedra generate each other[12]. Outside of these five solids, there were no others that could be made with equal-sided faces. It was only possible to use triangles (3-sided), squares (4-sided) and pentagons (5-sided). As we saw, using the hexagon (6-sided) as a face only produced a flat surface that nothing recognized from *the Flower of Life*.

THE FRUIT OF LIFE

Now, nothing saw that *the Fruit of Life* holds the two-dimensional projections of all five of the Platonic Solids.

Nothing began by fitting the cube's projection onto *the Fruit of Life*.

---

12. *the cube and the octahedron are known as each other's* duals. *The icosahedron and the dodecahedron are also each other's duals, and the tetrahedron is its own dual.*

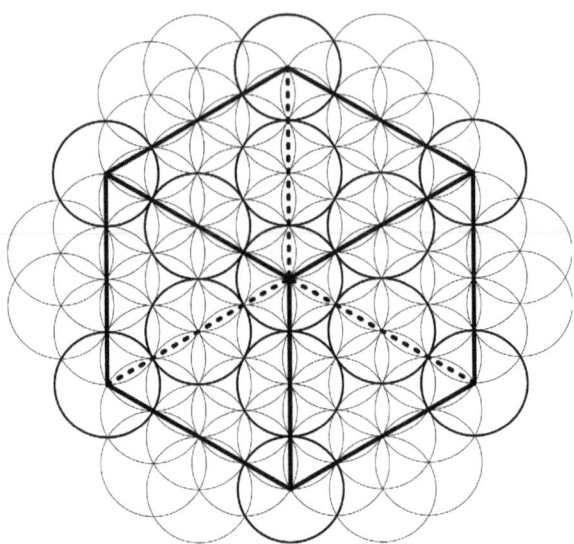

Then it recalled that the octahedron is held within the cube, and saw how perfectly it also fit into *the Fruit of Life*.

Now nothing saw that the tetrahedron also fits into the

cube, and shares four of its eight corners.

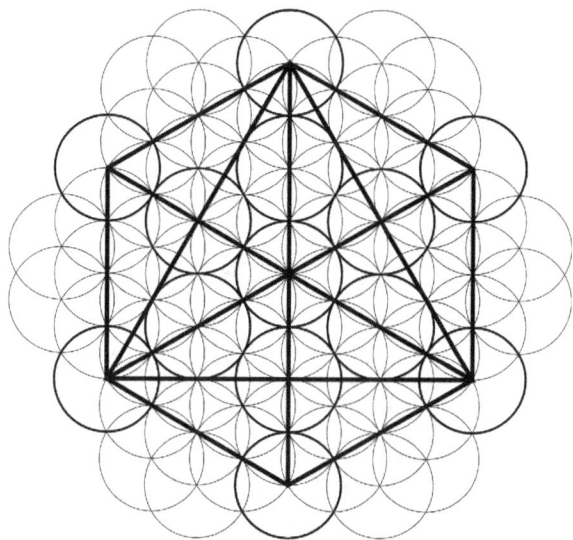

Next nothing projected the icosahedron onto the *Fruit of Life*, and saw that it was beautiful.

Remembering that the dodecahedron fits into the

icosahedron, nothing constructed the dodecahedron's projection onto *the Fruit of Life*.

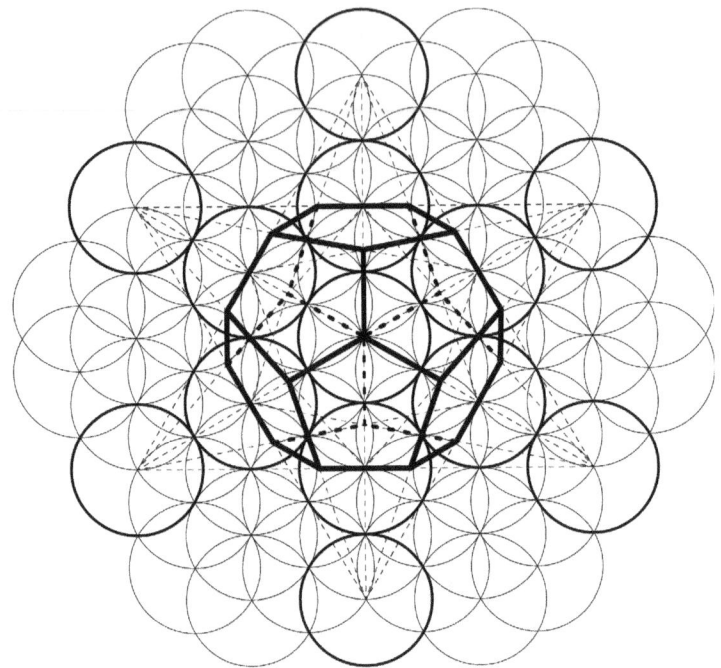

This we call *Metatron's Cube*[13] because the projections of all five Platonic Solids are contained within the projection of a cube on *the Fruit of Life*[14].

---

13. *Metatron is a Judaic archangel. In folkloristic tradition, he is the highest of the angels and serves as the celestial scribe or "recording angel". I do not know how this cube comes to take his name, but we can imagine the link.*

14. *This is the widespread claim in the tradition of Sacred Geometry, but technically the two-dimensional projections of the icosahedron and dodecahedron are here distorted to fit neatly onto the Fruit of Life. In actuality, their accurate two-dimensional projections do not coincide with the alignments in the Fruit of Life, and I will present my thesis on how their projections can be correctly constructed from the Fruit of Life in the Appendix.*

Nothing took a step back to behold the shapes and patterns that had already emerged soon after the Infinity-explosion, and it was pleased.

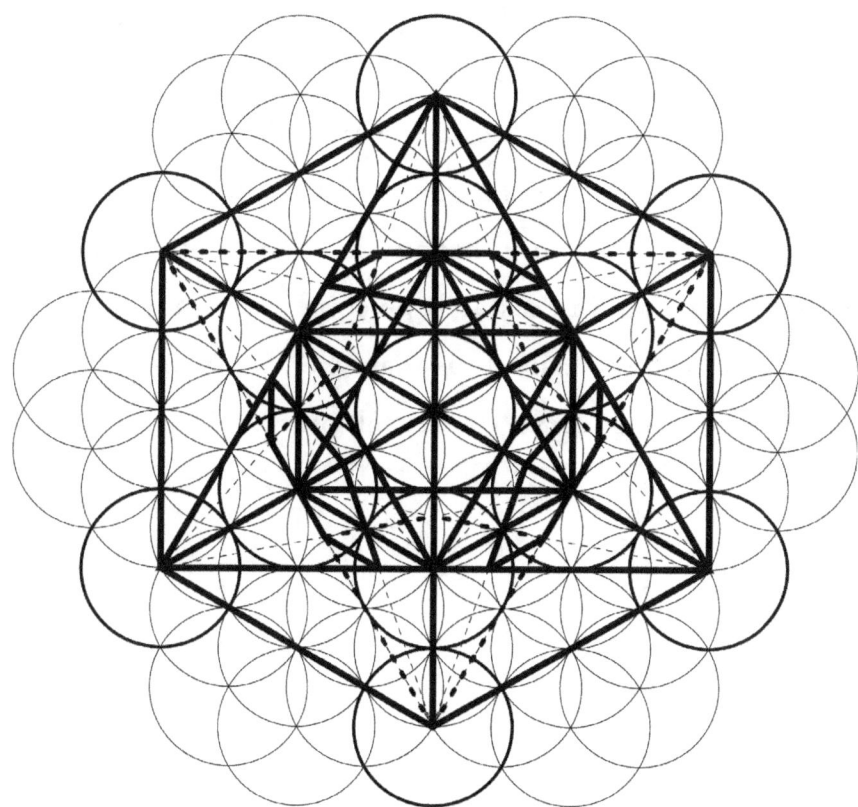

This was just the beginning of nothing's adventures, but now there was a beginning, and many more beginnings would follow. All of nothing's early adventures involved number and geometry because numbers and geometries are what emerge when you start with nothing—or zero—and infinity.

Nothing's native language was the language of mathematics—that language so shunned by most schoolchildren. But perhaps if they knew where this language came from, and who spoke it first, they would take a greater interest in it. It is also why scientists, as they try to uncover the secrets of our universe, inevitably find themselves speaking the language of mathematics. It would be reasonable to suppose that any form of intelligence in the universe would necessarily be conversant in mathematics. How marvelous that when we speak of nothing we can all understand each other.

APPENDIX

CHAPTER 3: Why Metatron's Cube is Wrong

When we looked at how the two-dimensional projections of all five Platonic Solids fit onto *the Fruit of Life* (See note 14. on page 34), we saw a neat arrangement that is in fact optically incorrect. While the projections of the tetrahedron, cube, and octahedron are isometrically[15] accurate, that of the dodecahedron and icosahedron only fit neatly onto the *Fruit of Life* if slightly distorted.

---

15. *Isometric projection is a method of visually representing a 3-dimensional object in two dimensions. In isometric projection, all three axes (length, width, and height) are projected onto a single plane, creating a distorted but still recognizable representation of the object.*

Unlike other types of projections, such as perspective projection, isometric projection does not attempt to create a realistic or accurate representation of the object's appearance. Instead, it aims to preserve the relative proportions of the object's features and show all three dimensions in a clear and understandable way.

To create an isometric projection, the object is first positioned in space at a particular angle relative to the viewer. Then, the object is projected onto the viewing plane by projecting each point on the object along lines that are parallel to the three axes of the object, but are tilted at a 45-degree angle to the viewing plane. This creates a distorted but proportional representation of the object that can be easily understood and analyzed.

Isometric projection is commonly used in technical drawing, engineering, and architecture to show the design of objects and buildings in a clear and concise way. It is also often used in video games and other forms of digital art to create stylized representations of 3D environments and objects.

Let's compare the distorted but neat projections with the optically accurate isometric projections. Here we see the icosahedron fitting onto *the Fruit of Life*:

Here we see the accurate isometric projection of the icosahedron onto *the Fruit of Life*:

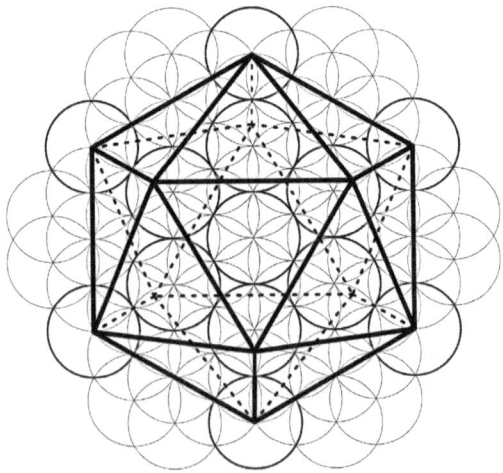

Here we see the dodecahedron fitting onto *the Fruit of Life*:

Here we see the accurate isometric projection of the dodecahedron onto *the Fruit of Life*:

While the differences between the projections are slight,

especially in the case of the dodecahedron, they do upset the neatness with which we would fit them onto *the Fruit of Life*. To see the differences more clearly, let's superimpose the projections:

This realization bothered me immensely, because I was certain that nothing would not cheat at geometry. Humans may cheat to idealize things, but nothing never cheats; it has nothing or no-one to deceive.

So I set out to solve this mystery, and tried to discover how nothing would have projected the five Platonic Solids onto *the Fruit of Life* accurately and correctly.

THE GOLDEN RATIO

To do so, we have to look at, and understand another of nothing's early discoveries.

Once nothing found that it could join points with straight lines, it realized that lines could be longer or shorter than one another. That is one way it discovered numbers, because the length of lines could be described by the number of equal-sized lines that were joined together.

If two equal lines were joined together, the new line would be twice (2) as long as the original (1) line. Similarly, if five equal lines were joined together, the new line would be five (5) times as long as the original (1) line.

Thus nothing came upon the *ratio*[16]. The ratio in this case is a comparison of the lengths of the two lines.

Now nothing wanted to create a beautiful and mysterious ratio. It wanted to divide a line such that the ratio of *the longer half* to *the shorter half* would be the same as the ratio of *the whole line* to *the longer half*.

In algebraic terms, if *the longer half* was of length=1, and if φ was *the shorter half*, nothing wanted

$$1 / φ = 1+φ / 1$$

Simplifying the equation, we end up with the quadratic equation,

$$φ2 + φ -1 = 0$$

whose positive solution is 0.618034

---

16. *A ratio is a comparison of two or more quantities that are related to each other. Ratios are typically expressed in the form of a fraction or a division expression, with the first quantity being the numerator and the second quantity being the denominator.*

For example, if there are 4 red balls and 6 blue balls in a bag, the ratio of red balls to blue balls is 4:6 or 4/6, which can be simplified to 2:3 or 2/3 by dividing both the numerator and denominator by their greatest common factor.

Ratios are commonly used to compare the relative sizes or amounts of two or more things. They can also be used to express probabilities, rates, or percentages. In mathematics, ratios are an important concept in many areas, including geometry, algebra, and statistics.

Nothing found that this mysterious ratio, known as *the Golden Ratio*, had already been waiting to be noticed in the pentagon, the shape of the face of the dodecahedron, as well as the shape of the base of five triangles joined around a point to build the icosahedron:

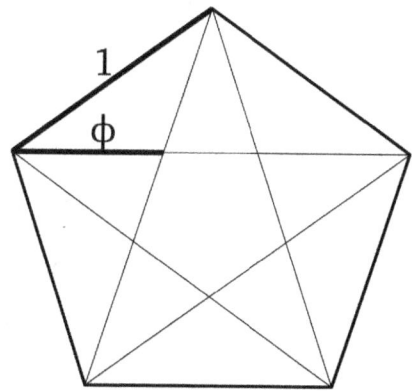

and even:

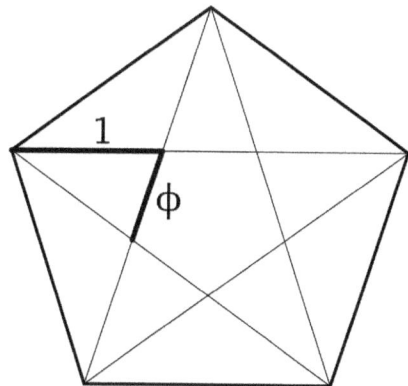

Thus nothing observed that the icosa- and dodeca-hedra had intimate connections to the Golden Ratio.

Metatron's Cube?

If the beauty of Metatron's Cube is that it describes a cube that holds the five Platonic Solids within it, and the ensemble projects onto *the Fruit of Life*, then why, I asked myself, does the cube neatly contain the tetrahedron and the octahedron as it would in three-dimensions, but the icosahedron and the dodecahedron float beyond and within the cube respectively?

If we consider that the isometric projections represent a three-dimensional situation, Metatron's Cube shows a cube with a tetrahedron nested within it, the tetrahedron's six edges forming diagonals on each of the cube's six faces; and with an octahedron nested within it, the octahedron's six points[17] touching the center of each of the cube's six faces. The icosahedron, however, protrudes beyond the cube, and the dodecahedron floats within it, without any significant relationship to the cube.

This felt unsatisfactory, especially given that in three-dimensions, the icosahedron and the dodecahedron can and do nestle neatly into a cube where the centers of each

---

17. *vertices*

of the cube's six faces are intersected by an edge of each of the two solids.

So it is worth taking a good and proper look at the isometric projections of how the tetrahedron, octahedron, icosahedron, and dodecahedron actually fit into the cube in three-dimensions[18].

Here we see the empty cube on its own:

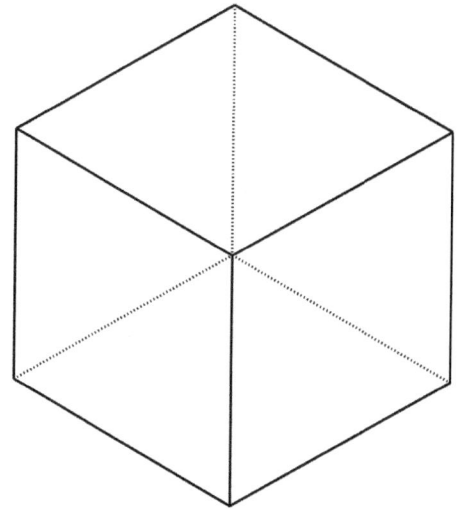

If we now place a tetrahedron whose sides are equal in length to the diagonals of the faces of the cube, we see that it fits exactly as it does in Metatron's Cube:

18. I produced these isometric projection views through accurate 3D modeling.

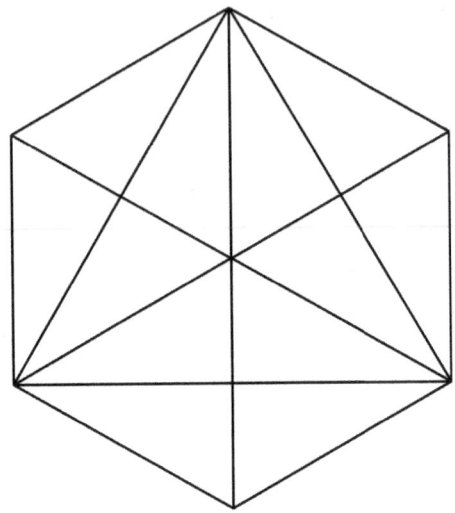

Removing the tetrahedron and replacing it with an octahedron whose vertex-to-vertex height is equal to the cube's, we see that it also fits exactly as it does in Metatron's Cube:

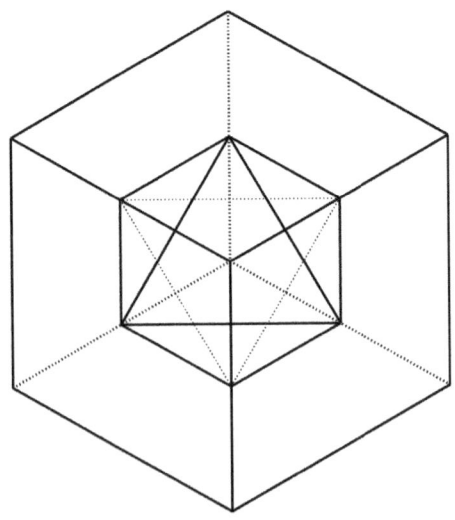

Now replacing the octahedron with an icosahedron, six

of whose sides lie flat along the cube's six faces we see

something quite different from Metatron's Cube:

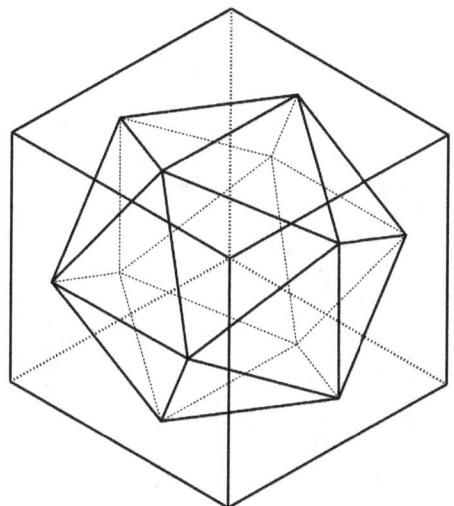

Likewise for the dodecahedron, this is not seen in

Metatron's cube:

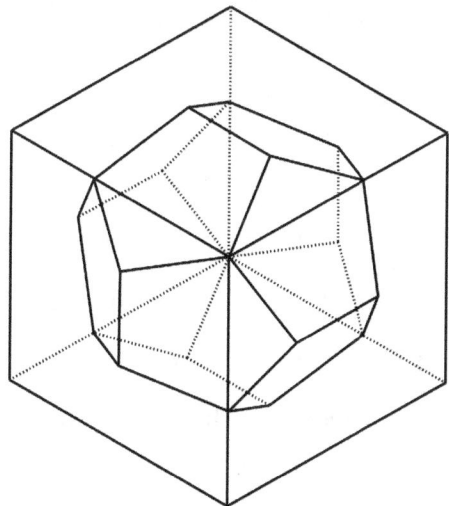

When we combine all solids together, neatly into the cube,

with optically accurate isometric projections, we find something quite beautiful, though of irregular symmetry:

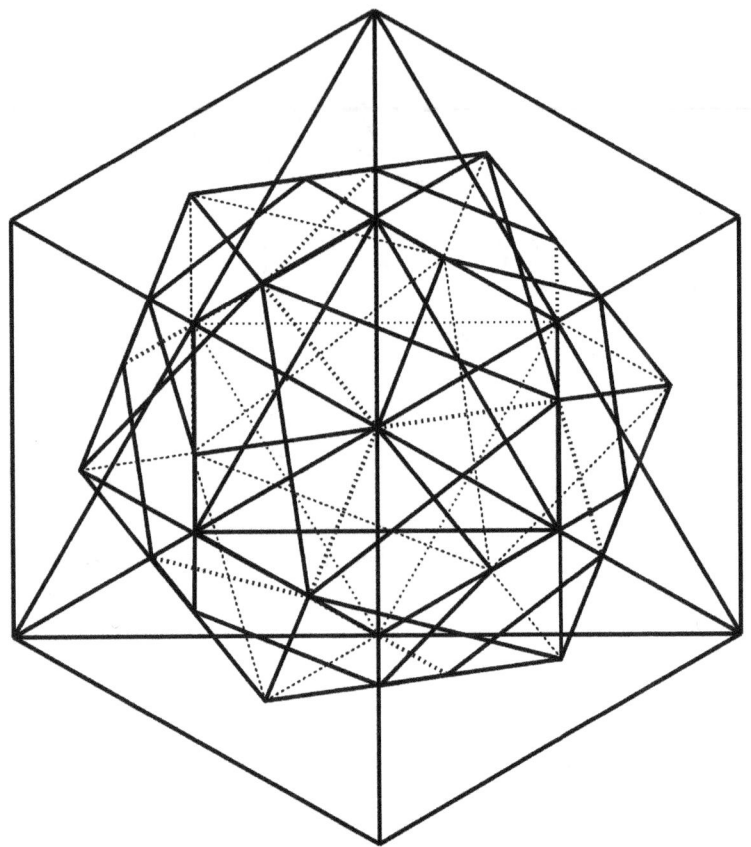

But how does this relate to *the Fruit of Life*? Well, the problem only arises with the representations of the icosahedron and the dodecahedron. Metatron's Cube already accurately represents the tetrahedron, cube, and octahedron against *the Fruit of Life*.

φ TO THE RESCUE

It is clear that the icosahedron and dodecahedron cannot neatly fit onto *the Fruit of Life* in this arrangement. So I asked myself whether instead *the Fruit of Life* could neatly fit beneath the icosahedron and dodecahedron in this isometric projection.

The answer was a beautiful "yes".

Given the prevalence of the Golden Ratio within the icosahedron and dodecahedron, I told myself that the answer must involve φ.

First let's look again at the icosahedron in Metatron's Cube, set against *the Fruit of Life*:

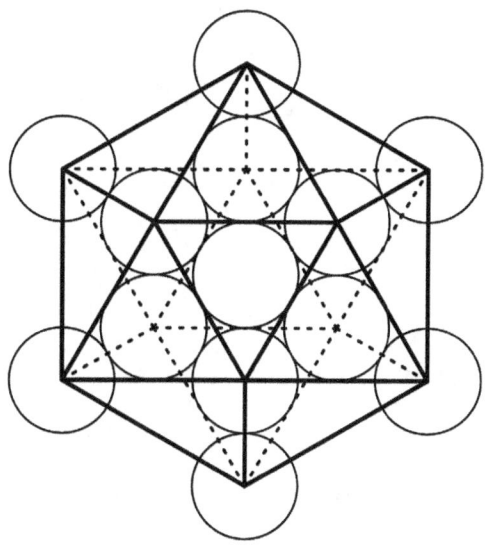

We can observe that the icosahedron's outline is a hexagon whose vertices coincide with the centers of the third ring of circles from the center. The icosahedron's six other vertices coincide with the centers of the second ring of circles from the center.

Looking at the accurate isometric projection of the icosahedron against *the Fruit of Life,*

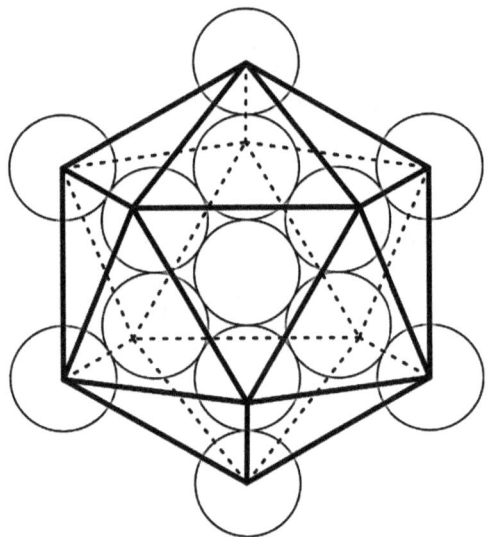

we see that the vertices of the icosahedron's outline remain unchanged and coincide with the centers of the third ring of circles from the center. However the icosahedron's remaining vertices no longer coincide with any feature on *the Fruit of Life.*

Now, if we take *the Fruit of Life,*

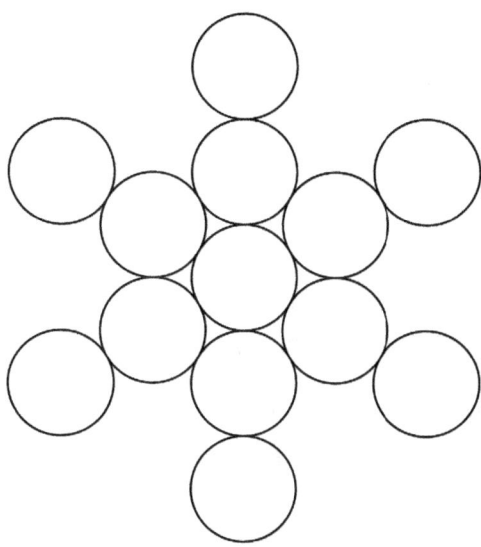

and duplicate it, then multiply it by the Golden Ratio, we get two *Fruits of Life,* the original at 100% size and a smaller one at 61.8034% size:

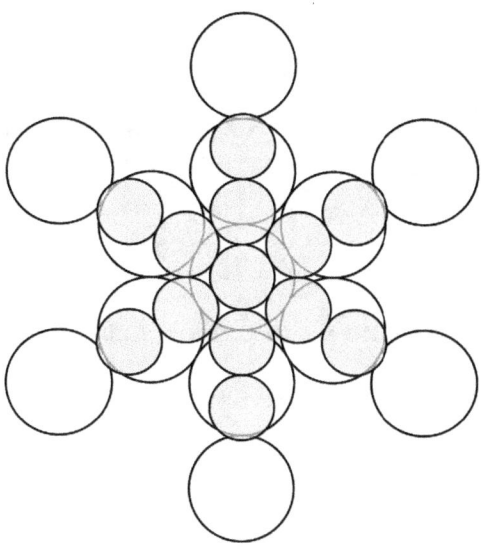

I will refer to the original *Fruit of Life* at 100% size as *the Fruit of Life (1)* and the smaller *Fruit of Life* at 61.8034% size as *the Fruit of Life (ϕ)*.

Projecting the isometrically correct icosahedron onto the *Fruits of Life*, we find that the icosahedron's vertices all coincide exactly with the centers of the circles: its outer vertices with those of *the Fruit of Life (1)*, and its inner vertices with those of *the Fruit of Life (ϕ)*. What a beautiful and meaningful alignment!

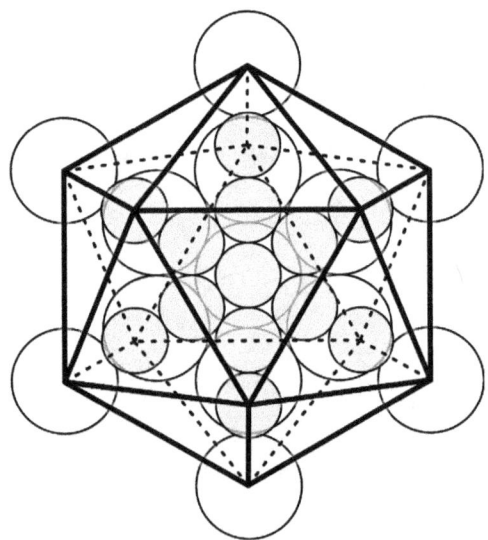

Thus we see that the icosahedron's accurate isometric projection is consistent with *the Fruit of Life* and the Golden Ratio.

A closer look at it against *the Fruit of Life (ϕ)* reveals a beautiful geometrical order at work.

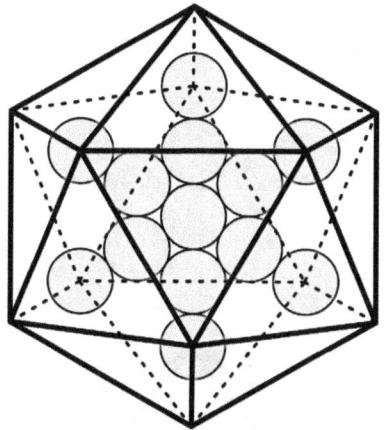

This takes care of the undistorted isometric projection of the icosahedron. But now we must account for the correct isometric projection of the icosahedron nested within the cube in terms of *the Fruit of Life*, i.e.:

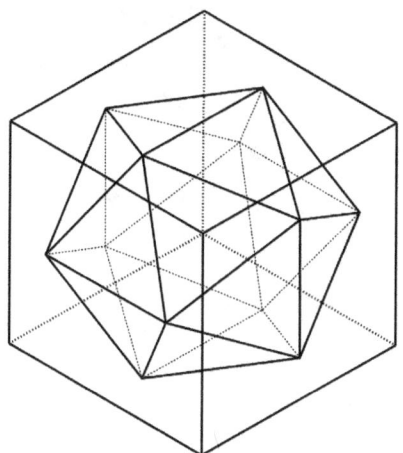

It appears that at least two transformations are necessary to arrive at this final, accurate projection. Relative to the cube's projection, the icosahedron's projection is 1) scaled down by an unknown factor, and 2) rotated by an unknown angle.

We begin with the icosahedron superimposed upon the cube on *the Fruit of Life (1)*, as in Metatron's Cube:

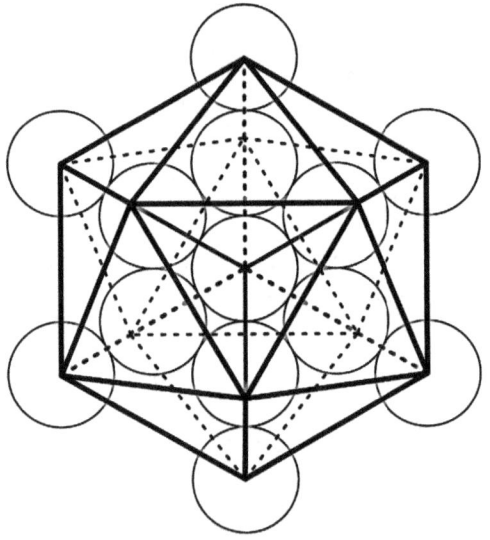

THE GOLDEN TRANSFORMATION

I speculated that we would need a transformation that satisfies the Golden Proportion, and explored the following:

"How do I both scale and rotate a hexagon (*hexagon'*), such that the vertices of the resulting hexagon (*hexagon"*) touch the edges of *hexagon'* at a points that divide the edges of *hexagon'* according to the Golden Ratio?"

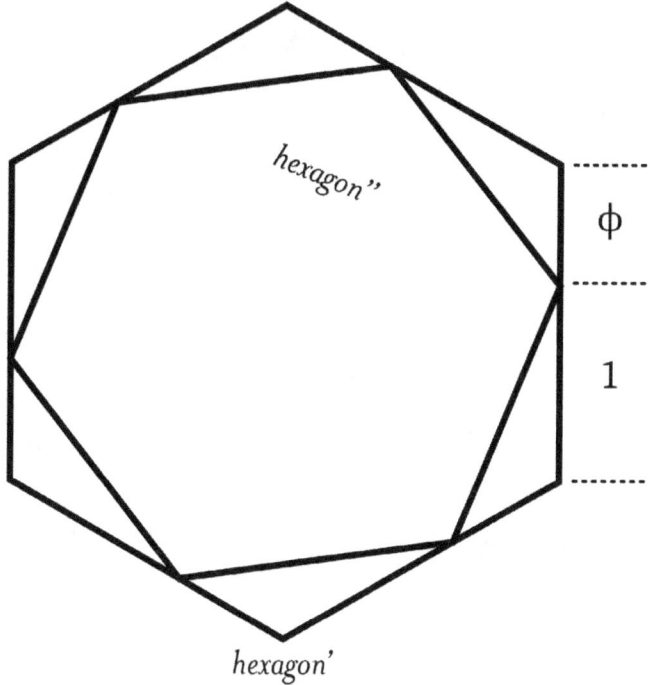

hexagon'

The answer was to be found in the following calculations:

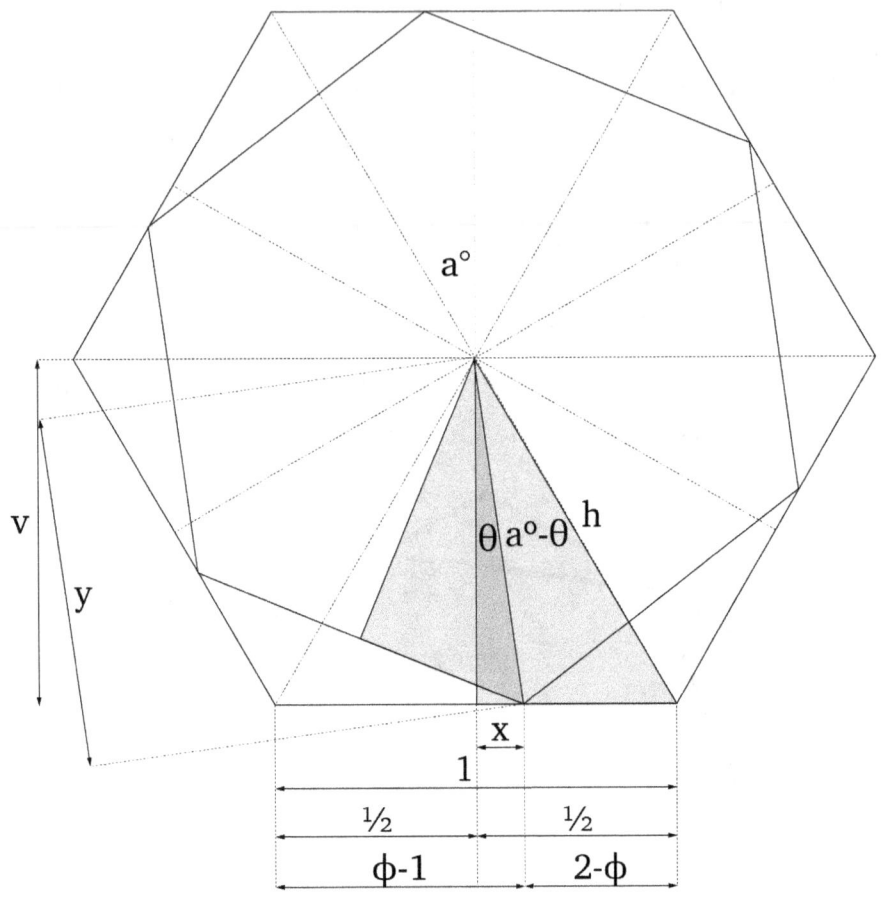

angle between bissected edge and bissected vertex

a = (360/2n) in °

length (hypotenuse) of bissected vertex

h = 1/2 * sin(a)

perpendicular distance from center to edge

$v = h*cos(a)$

displacement from midpoint in golden proportion

$x = \phi - 3/2$

$\theta = atan(x/v)$ in °

$rotation = (a) - (\theta)$

$y = x/sin(\theta)$

$scale = y/h$ in %

n number of sides of polygon

n=6

a=> 30°

h=> 1

v=> 0.8660254038

θ=> 7.761243907°

rotation=> 22.238756093°

y=> 0.8740320489

scale=> 87.4032048898%

Scaling the icosahedron by 0.8740 and rotating it by 22.2388°, we see that it finds its correct orientation within the cube, but remains too large to be nested within the cube in three-dimensions, and loses its correspondence to *the Fruit of Life*.

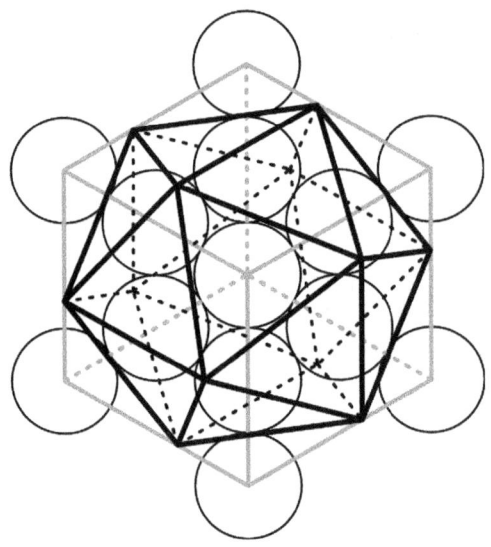

So we apply the same transformations to *the Fruit of Life* (1), and create a third *Fruit of Life* that has been transformed in accordance with the Golden Ratio, which we will call *the Fruit of Life* (1'):

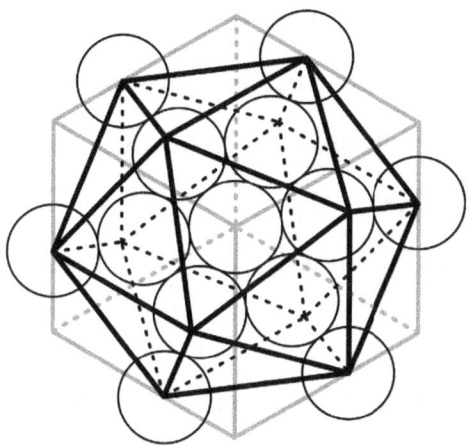

Similarly, we can apply the same transformations to *the Fruit of Life (ϕ)* and obtain *the Fruit of Life (ϕ')*:

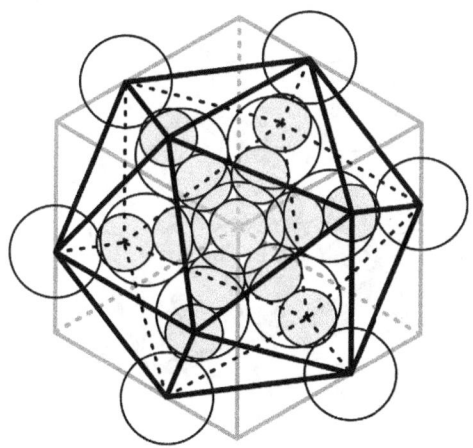

Speculatively, let us shrink the icosahedron so that its inner vertices shift from the centers of the third ring of circles on *the Fruit of Life (ϕ')* to the centers of the second ring of circles on *the Fruit of Life (1')*.

In order to calculate this scaling factor, we observe that the centers of the third ring of circles on *the Fruit of Life* (ϕ') are 4*ϕ units away from the absolute center, and the centers of the second ring of circles on *the Fruit of Life* (1′) are 2*1 units away from the absolute center.

Thus the scaling factor is 1/2ϕ which is 0.8090. Applying this to the icosahedron, we find that it assumes exactly the right dimensions of its correct isometric projection nested within the cube, while neatly enclosing *the Fruit of Life* (ϕ')!

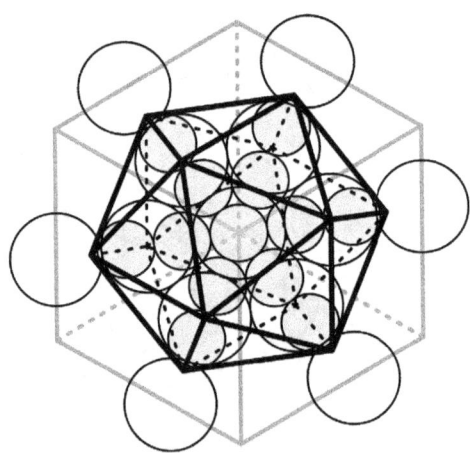

We can further observe that the icosahedron's outer edges now exactly intersect the centers of the third ring of circles of *he Fruit of Life* (ϕ):

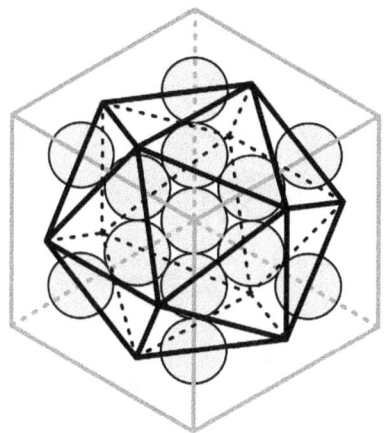

Taken all together, we have exact correspondences between the isometrically correct icosahedron and four *Fruits of Life* in Golden Ratios to each other *(1)*, *(1')*, *(φ), and (φ')*.

The four *Fruits of Life* in Golden Ratios to each other *(1)*,

(1'), (φ), and (φ') form a beautiful lattice upon which all Five Platonic Solids nested within the cube in three-dimensions are isometrically correctly represented. The lattice itself deserves contemplation:

Taken as the icosahedron's dual, the dodecahedron undergoes exactly the same transformations to achieve the same correct result.

The dodecahedron within the cube against *the Fruit of Life* *(1)*:

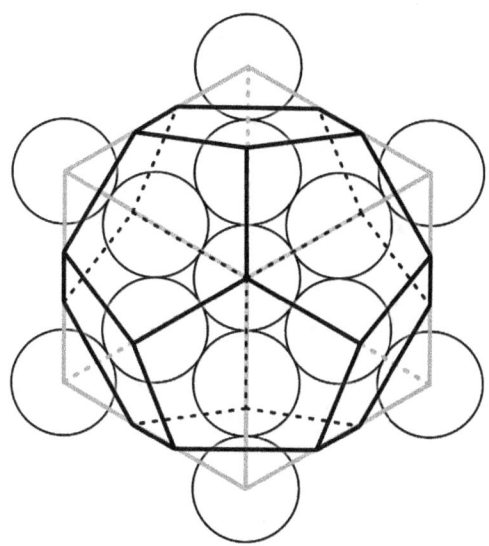

The dodecahedron against *the Fruit of Life* (ϕ):

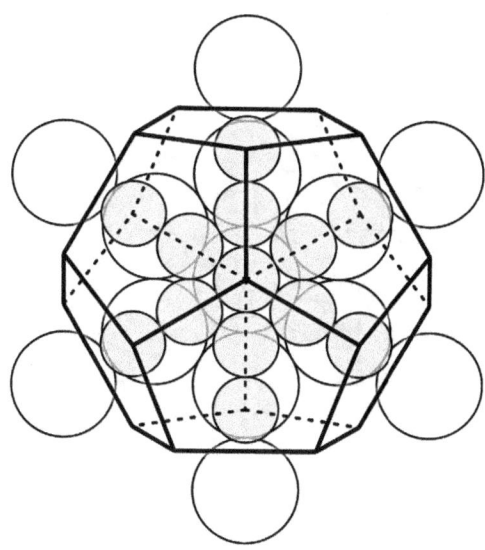

The dodecahedron scaled and rotated according to the *Golden Transformation*:

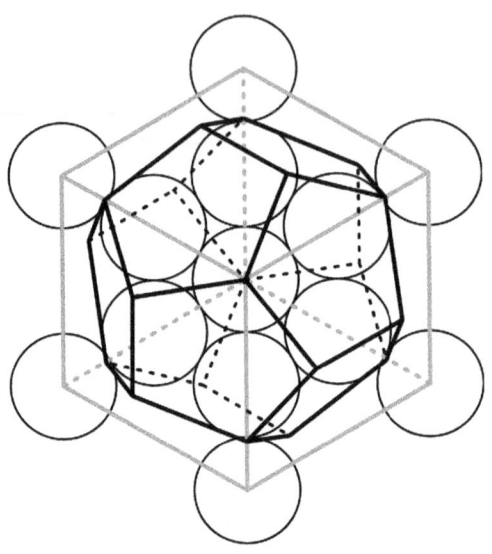

The scaled and rotated dodecahedron against *the Fruit of Life (1')* and *the Fruit of Life (φ')*:

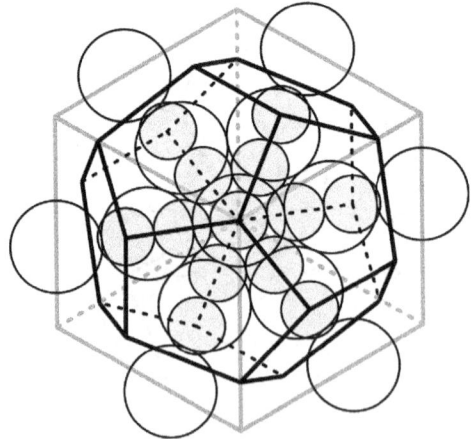

The scaled, rotated, and scaled dodecahedron against *the*

Fruit of Life (1') and the Fruit of Life (φ)

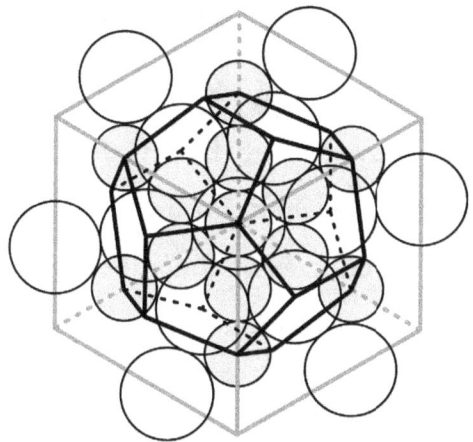

Taken all together, we have exact correspondences between the isometrically correct dodecahedron and four *Fruits of Life* in Golden Ratios to each other *(1), (1'), (φ), and (φ')*.

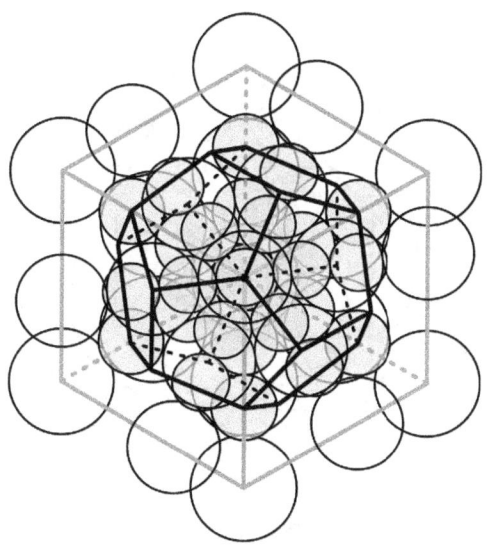

HENTSCH CUBE

So I daresay that my cube has the following properties:

In three-dimensions, it holds the four other Platonic Solids neatly whereby each of the tetrahedron, icosahedron, and dodecahedron have six edges that centrally touch the six faces of the cube, and the octahedron's six vertices touch the centers of the cube's six faces.

The isometric projection of the ensemble has exact correspondences to the centers of the circles of the *Fruits of Life*, where there are four *Fruits of Life* in Golden Ratios to each other *(1)*, *(1')*, *(φ)*, and *(φ')*.

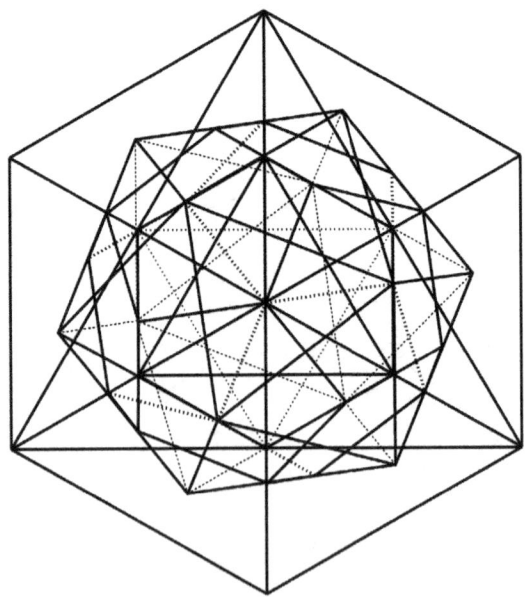

THE HENTSCH CUBE:

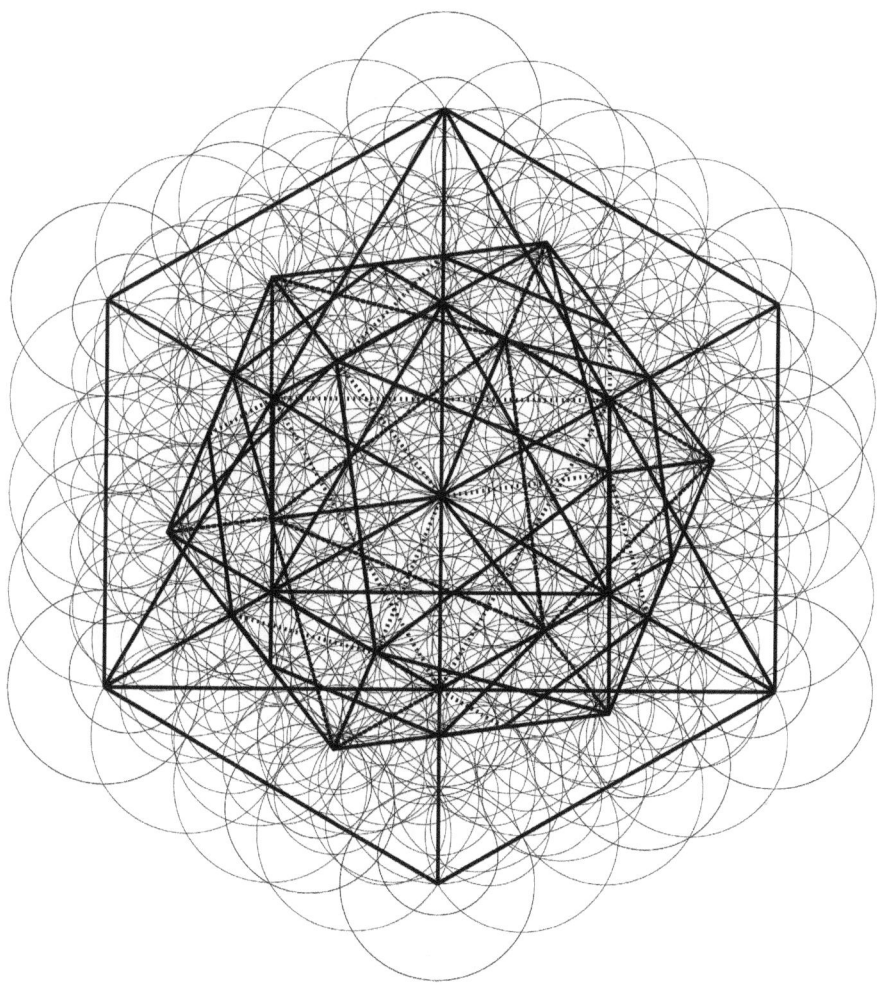

Let's look at the lattice made by the four *Fruits of Life* in Golden Ratios to each other *(1)*, *(1')*, *(ϕ)*, *and (ϕ')*, that I will call *the Golden Lattice*:

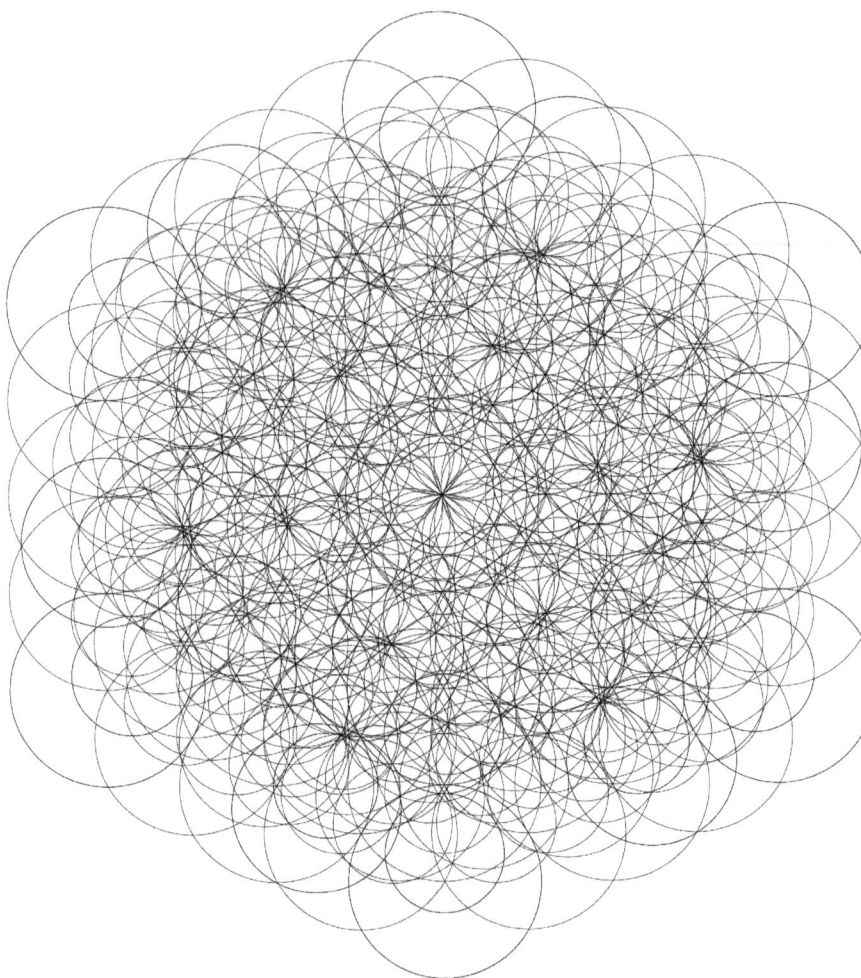

We can observe that there is a fuzzy indication—in the interference pattern—of where the vertices of the Hentsch Cube lie, while the vertices correspond exactly to the centers of the relevant circles in the *Golden Lattice*.

Contemplating the results, let's examine each of the five Platonic Solids against the *Golden Lattice*.

The Tetrahedron

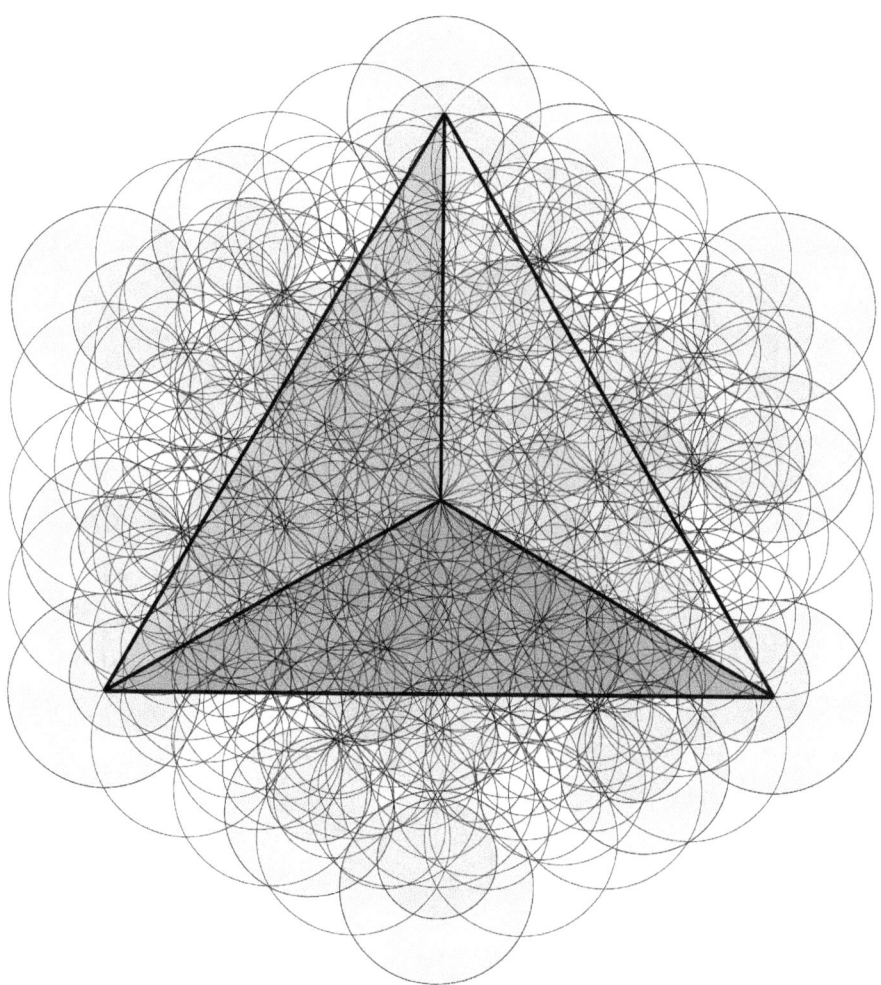

THE CUBE

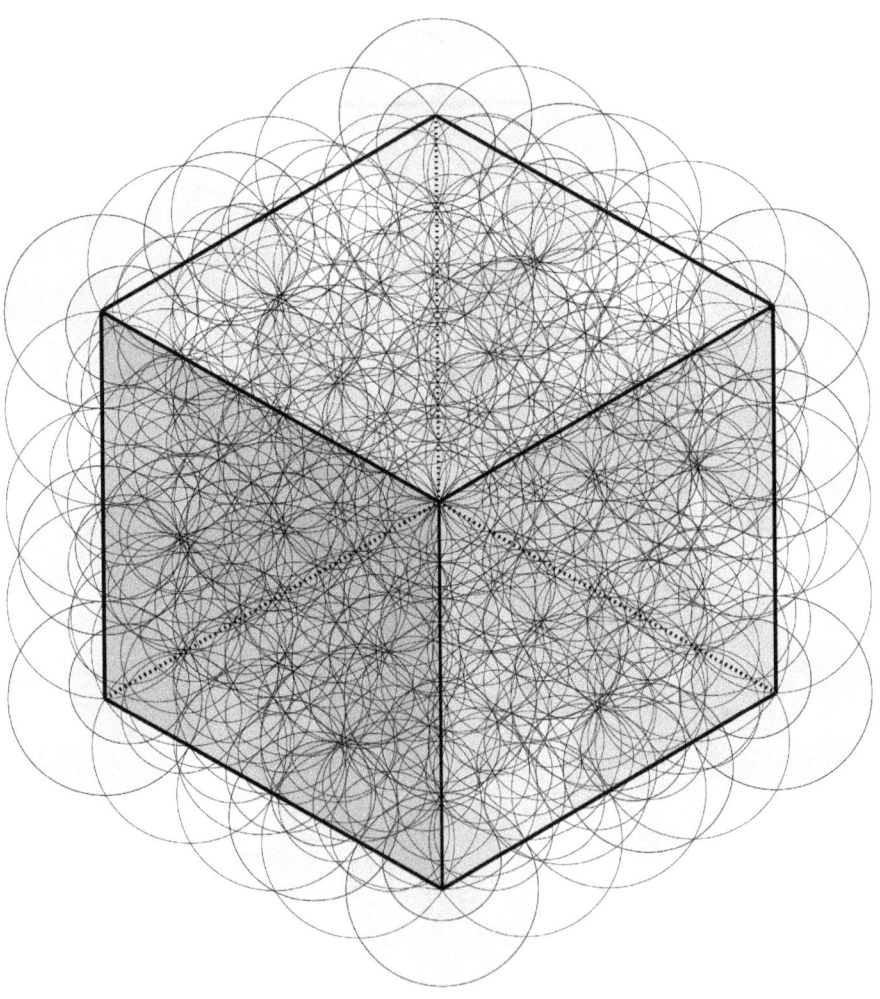

The Octahedron

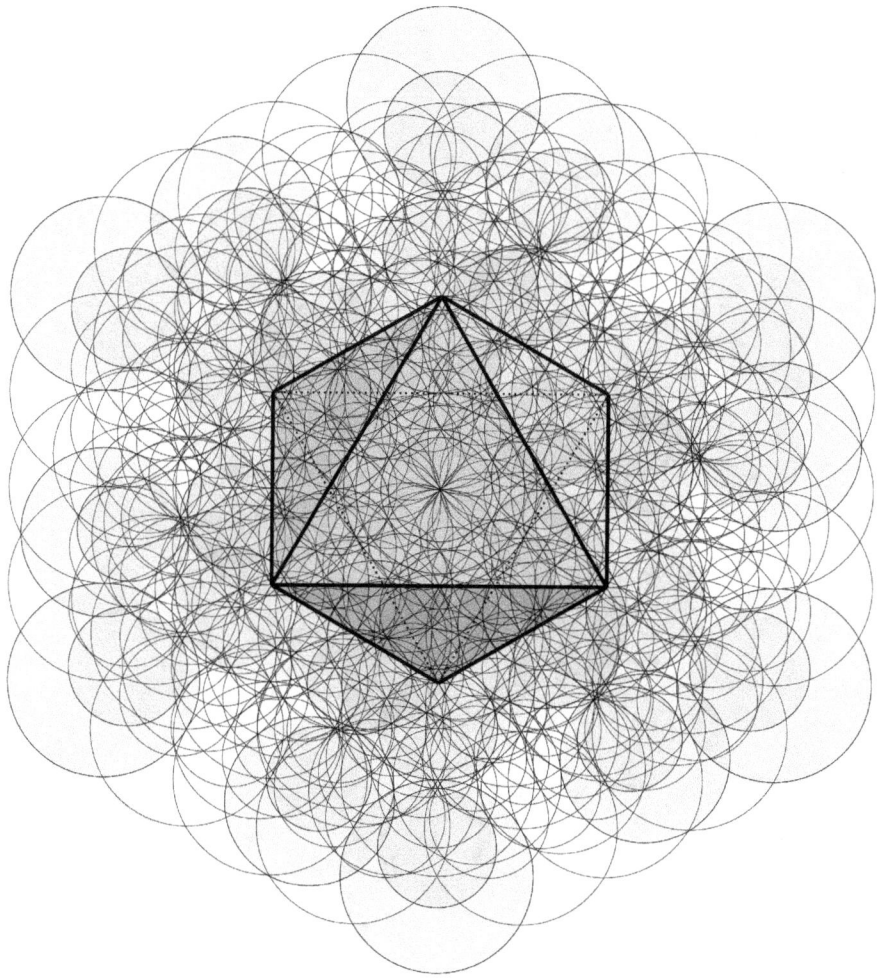

THE DODECAHEDRON

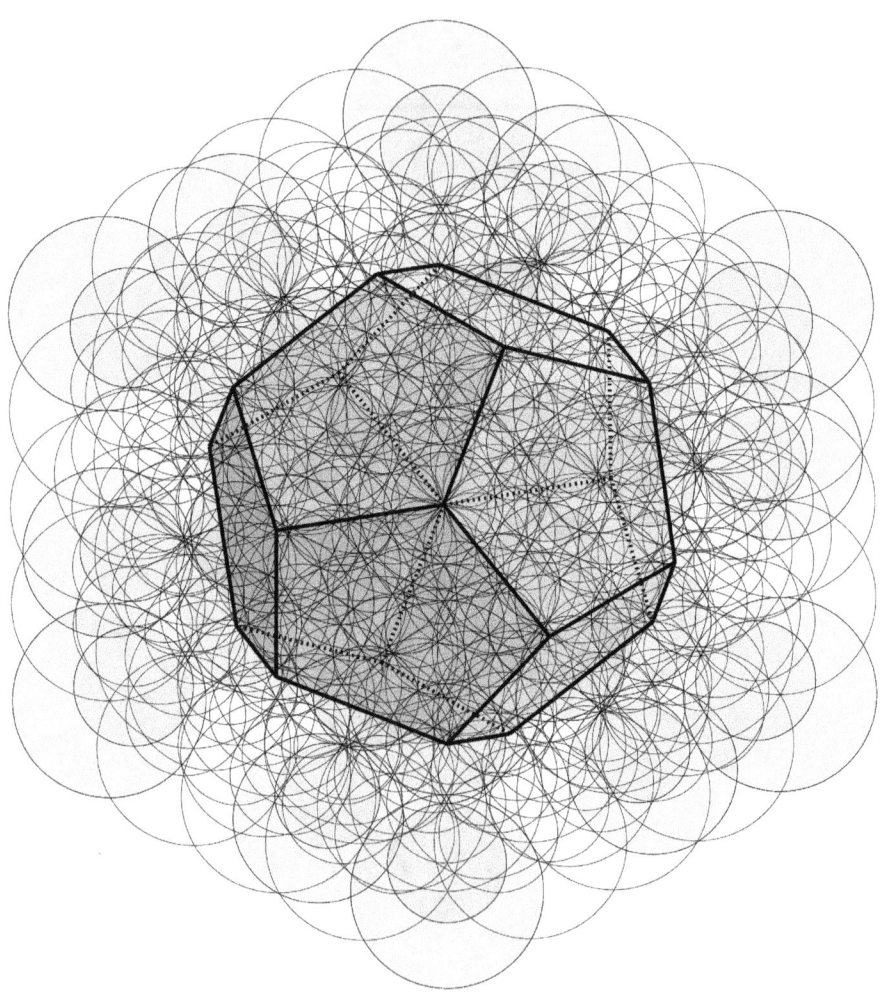

THE ICOSAHEDRON

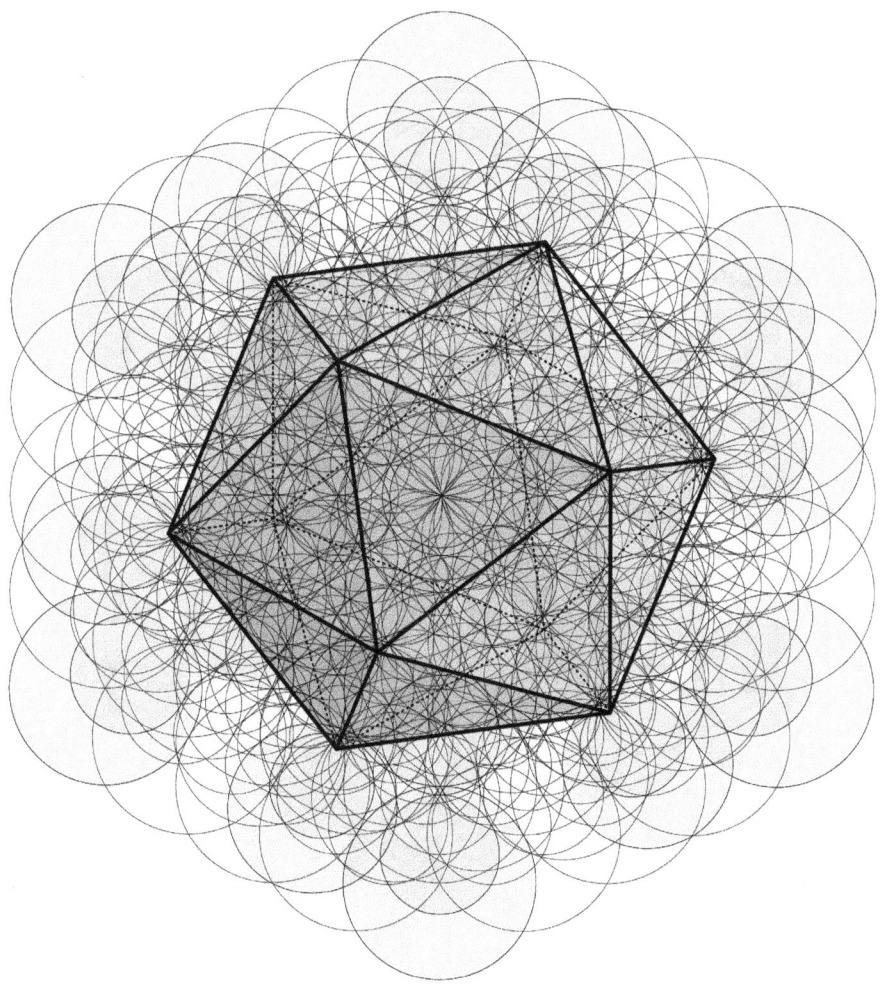

THE TETRAHEDRON WITHIN THE CUBE

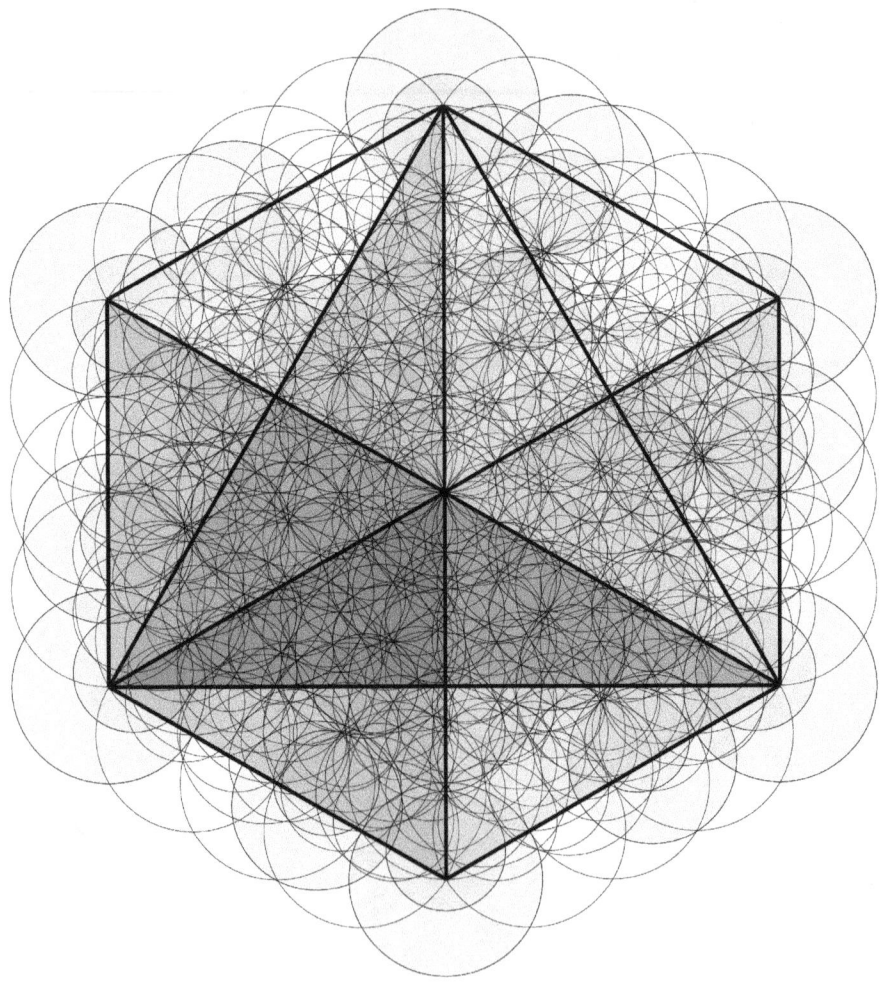

The Octahedron Within the Cube

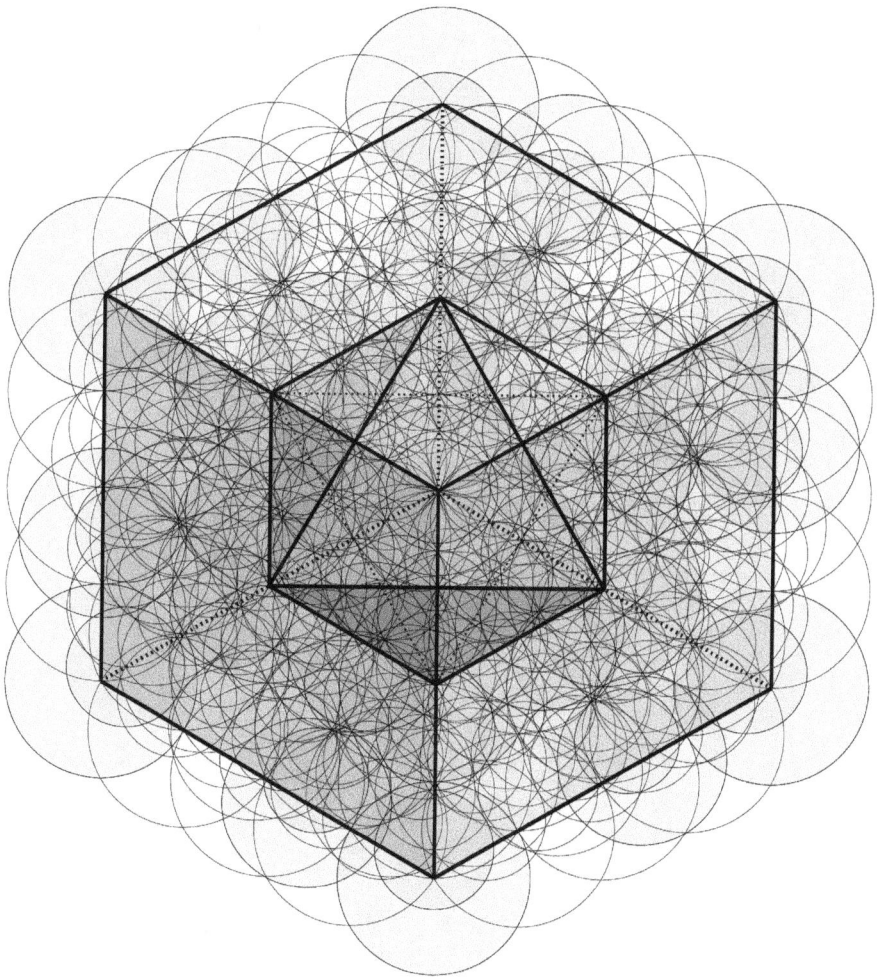

THE DODECAHEDRON WITHIN THE CUBE

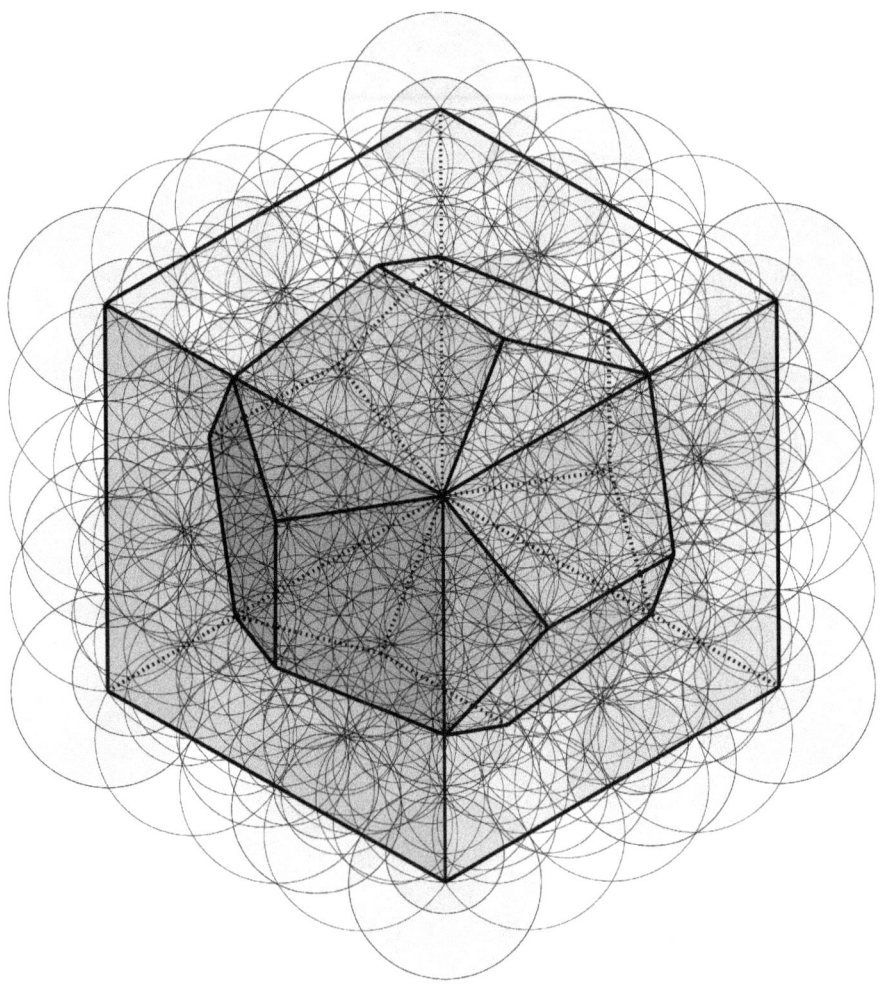

THE ICOSAHEDRON WITHIN THE CUBE

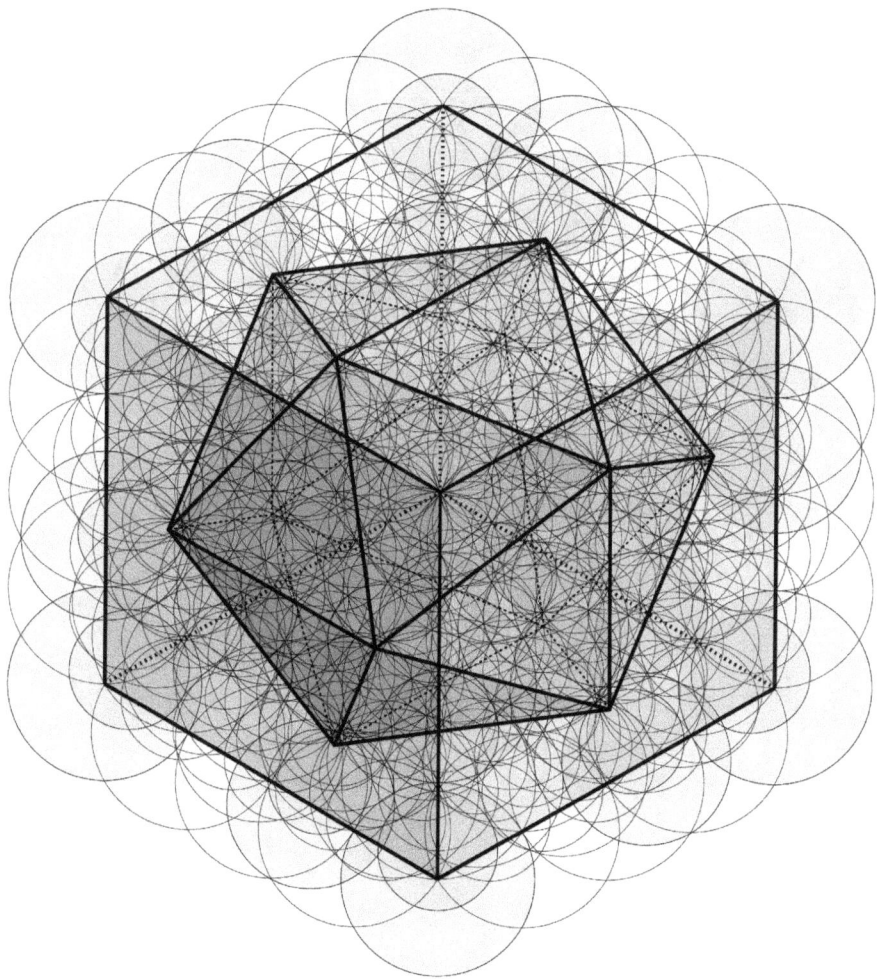

THE GOLDEN LATTICE

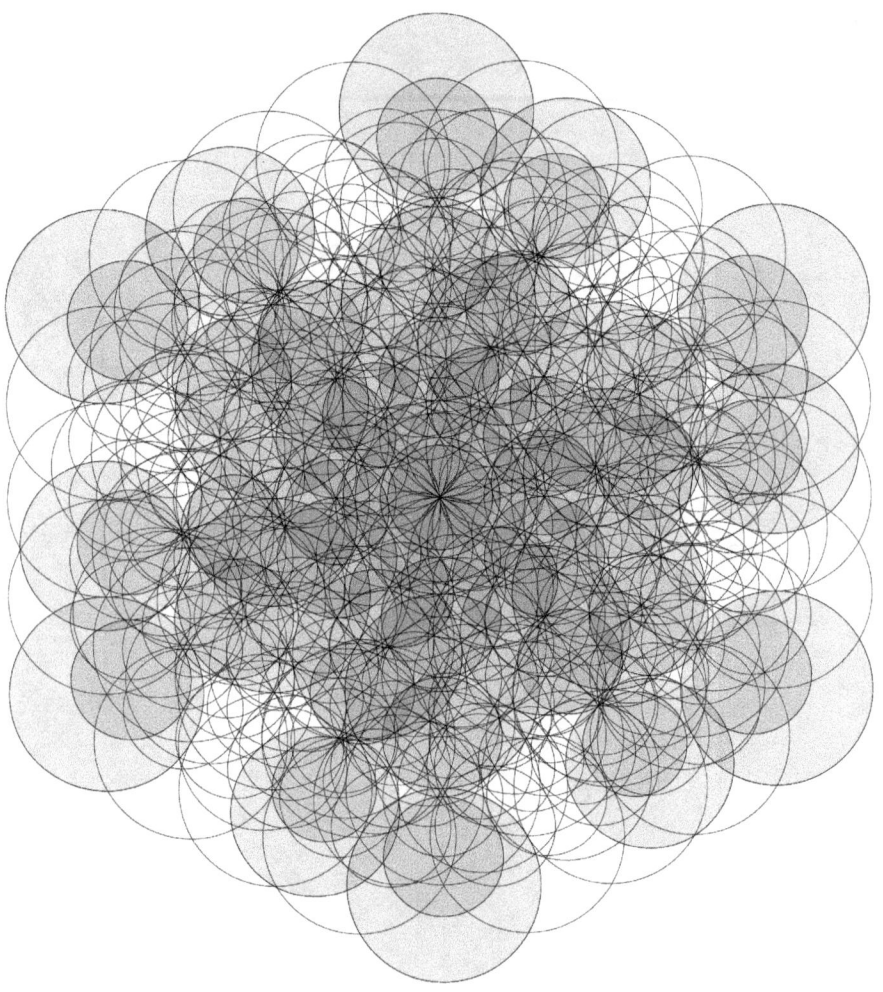

THE HENTSCH CUBE

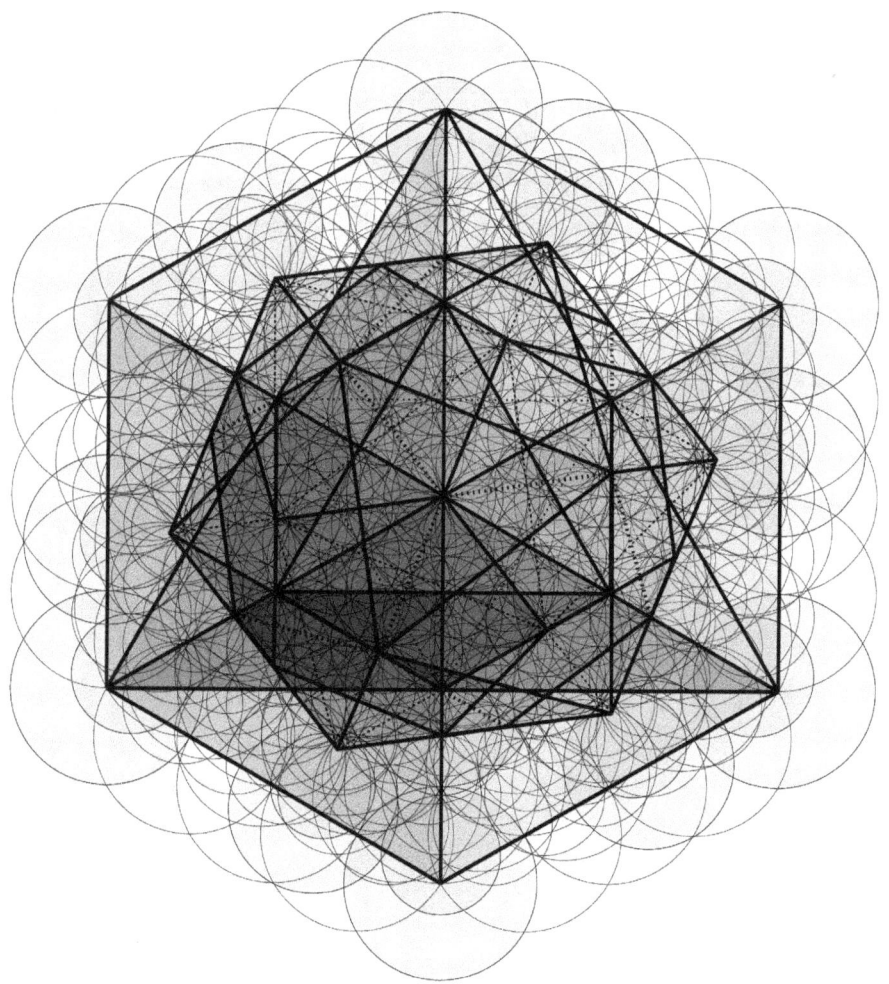

ACKNOWLEDGEMENT

I thank *nothing* for this work. I therefore thank everything and everyone. This includes computers, without which I would never have had the patience to explore the gap between the second and the third dimensions.

www.ingramcontent.com/pod-product-compliance
Lightning Source LLC
Chambersburg PA
CBHW072150230526
45467CB00042B/1511